Pervasive Information Architecture

Pervasive Information Architecture: Designing Cross-Channel User Experiences

Andrea Resmini and Luca Rosati

AMSTERDAM • BOSTON • HEIDELBERG • LONDON
NEW YORK • OXFORD • PARIS • SAN DIEGO
SAN FRANCISCO • SINGAPORE • SYDNEY • TOKYO

Morgan Kaufmann Publishers is an imprint of Elsevier

Acquiring Editor: Rachel Roumeliotis
Development Editor: David Bevans
Project Manager: Andre Cuello
Designer: Alisa Andreola

Morgan Kaufmann is an imprint of Elsevier
30 Corporate Drive, Suite 400, Burlington, MA 01803, USA

Library of Congress Cataloging-in-Publication Data
Resmini, Andrea.
Pervasive information architecture : designing cross-channel user experiences / Andrea Resmini, Luca Rosati.
    p. cm.
  Includes bibliographical references.
  ISBN 978-0-12-382094-5 (pbk.)
  1.   Ubiquitous computing. 2.   User interfaces (Computer systems) 3.   Information technology. 4.   Information storage and retrieval systems–Architecture. 5.   Cross-platform software development. I. Rosati, Luca, 1971- II. Title.
QA76.5915.R47 2011
005.1–dc22

2010053834

British Library Cataloguing-in-Publication Data
A catalogue record for this book is available from the British Library.

ISBN: 978-0-12-382094-5

Contents

Contributors

Kars Alfrink   Freelance
Gianni Bellisario   Rai - Radio televisione italiana
Cennydd Bowles   Clearleft
Stefano Bussolon   University of Trento
Terence Fenn   University of Johannesburg
Chiara Ferrigno   Rai - Radio televisione italiana
Claudio Gnoli   University of Pavia
Andrew Hinton   Macquarium
Jason Hobbs   Human Experience Design
Peter Morville   Semantic Studios
Eric Reiss   FatDUX
Jonas Söderström   inUse
Donna Spencer   Freelance
Samantha Starmer   REI

Foreword

I knew the Internet before it got famous. There were places but no paths, no maps, no search engines. Entry required a key in the form of an IP address and an incantation in the language of UNIX. It was a small world that felt big because it was so easy to get lost in the shadowy realm of texts and data, completely devoid of color. And yet, the Internet in the early 1990s was a friendly place because it was also made up of people who served as mentors and guides, helping one another find their way. This pioneering community of geeks and wizards, teachers and students, scientists and librarians was radically global and breathtaking in its diversity, and yet there was one thing we all held in common: a fervent belief that the Internet was about to change the world.

Now I've got that feeling once again. But this time there's no single protocol or portal to point to as evidence of what's to come. We're creating multichannel, cross-platform, transmedia, physicodigital user experiences that tear down the walls between categories. We can call it ubiquitous computing, the Internet of Objects, Web Cubed, or the Intertwingularity. We can talk about smart things, sensor Webs, product-service systems, and collaborative consumption. But none of these labels begins to describe the extraordinary diversity of the ambient, pervasive, mobile, social, real-time mashups unfolding before our very eyes. No word or phrase can possibly bind together the 21st-century success stories of iTunes, Nike+, Netflix, Redbox, Zipcar, iRobot, Freecycle, and CouchSurfing with the emergent phenomena of augmented reality, urban informatics, and plants that tweet. But as we wander blindly in this landscape of vernacular chaos, one thing is clear: we need a new map.

In 1998, Louis Rosenfeld and I coauthored the first edition of *Information Architecture for the World Wide Web*. The "polar bear book" as it came to be known helped a generation of information architects and user experience designers make sense of the Web through structure, organization, navigation, and search. Today, much of what we wrote remains relevant, and yet new questions arise as the world and the Web intertwingle. How will we decide which features belong on which platforms? How should we strike a balance between cross-channel consistency and platform-specific optimization? How do we rise to the new challenges of creating paths and places that bridge physical, digital, and cognitive spaces?

That's why I'm so excited by *Pervasive Information Architecture* and the heroic efforts of Andrea Resmini and Luca Rosati to explore, uncover, and chart the new, new world that's surrounding us all. This refreshing book about the design of ecosystems for wayfinding and understanding promotes a holistic approach to information architecture and user experience that draws insights from multiple disciplines and historical contexts. And it leads us bravely into the future with an ingenious collection of medium-independent heuristics to guide the complex decisions that lie ahead. In short, Andrea and Luca have sketched a map to the future of cross-channel design that will in turn inspire the next generation of mapmakers to improve usability, findability, and desirability and to make the world/Web a better place.

Peter Morville
Ann Arbor, Michigan
2011

Attribution

This book is the result of a collaborative effort: the authors have discussed together each aspect of it. However, Andrea Resmini has written Chapters 2, 3, 4, 8, 9 (first half) and revised Luca's work; Luca Rosati has written Chapters 1, 5, 6, 7, 9 (second half) and revised Andrea's.

Introduction

FIGURE 1
Photo: B. Ostrowsky.
Source: Flickr.

PERVASIVE

"But I like maps. I've got maps all over my house. I'm going to suggest
to you that the skills and knowledge we have all been developing in
our work—especially pertaining to the Internet—have application out
here." He taps the whiteboard. "In the real world. You know, the big
round wet ball where billions of people live"

(Stephenson 1999).

This is a book on design.

This is also a book on information architecture, as we research and practice
information architecture as a design proposition, but it's not about Web sites.
Or better, it's not only about Web sites. The reason for this is because the way
we interact with information is changing.

Imagine a pendulum. It swings right. Then it swings back. And that's what
we are doing, we are swinging back. Fifty years ago, if you wanted to know
something, you asked a friend, a teacher, your parents, your siblings, or your
spouse. Or you read a book. Physical entities. Then, technology made the desktop computer our interface of choice to access information, with a dedicated

place in the house, the office, and the school. Now we are swinging back to the real world, but we are bringing the computers along, and they are restless, smaller, faster, connected.

Information is going everywhere. It is bleeding out of the Internet and out of personal computers, and it is being embedded into the real world. Mobile devices, networked resources, and real-time information systems are making our interactions with information constant and ubiquitous. Information is becoming pervasive.

More and more of what we do every day requires us to move among different media, channels, and environments, with no distinction between what is physical and what is digital. We still visit Web sites, but we also use mobile applications, interact with intelligent devices, and connect with people through a variety of computer-mediated technologies. And we move on and off: check something out on the Web, get updates via text messages, go to the shop, bring the thing home, use it, connect it, or get more services online.

We call these new sprawling information spaces–in which we interact with both digital and physical entities–ubiquitous ecologies: they are systems connecting people, information, processes, and they are everywhere. They are pervasive information architectures. They are the structuring layer that runs across the different media, channels, and processes in which we express our expanded self, socially. The boundaries are blurring.

How do these changes affect the design of these information spaces then? Can we still be satisfied with designing our Web site, our mobile application, our customer-facing way-finding signage, or our kiosk system, in splendid isolation? Single, fully-realized artifacts we offer for use? We believe not.

We believe that we have to start designing these as the seamless, cross-channel holistic journeys we are experiencing them as. Traveling, shopping, taking care of our health, or enjoying ourselves: even when we design but a small part of these, as it often happens, knowing it does not end there, with just a ticket, an online shopping cart, or some info received on a mobile phone. Knowing that our loose piece is part of a larger, complex ubiquitous ecology: that is going to make the difference. Changes everything, actually.

STRUCTURE OF THE BOOK

The book is divided into three parts. Because we like big words that require a dictionary just to be spelled correctly, we didn't call these Parts 1, 2, and 3 originally. That would have been too easy. We called them Foundations, Heuristics, and Synthesis, which is a bit like saying "A few facts and knowledge we rely on" (Foundations), "Guidelines we use when designing" (Heuristics), and "Putting it all together" (Synthesis), but in fewer words. You never know. Our editors suggested that we could compromise: they could be persuaded into leaving those names around if we added Parts 1, 2 and 3 somewhere so that readers could understand that they were dealing with different parts of the book even without a dictionary. We agreed; it seemed wise.

These parts have a different structure and serve different purposes:

- Part 1, *foundations*, simply introduces the book, provides an overview of the problem space, and offers a historical read of information architecture as both a field of practice and a research discipline. Part 1 contains this *Introduction* and two more chapters, *From Multichannel to Cross-channel* and *Toward a Pervasive Information Architecture*.
- Part 2, *heuristics*, provides an initial overview of what designing pervasive information architectures means, what it tries to address, and illustrates the five guiding design principles of place-making, consistency, resilience, reduction, and correlation in detail. Part 2 deals mostly with the conceptual model and with its practical impact on the design process when dealing with cross-channel user experience and contains six chapters: *Heuristics for a Pervasive Information Architecture, Place-making, Consistency, Resilience, Reduction,* and *Correlation.*
- Part 3, *synthesis*, recomposes these principles or heuristics into a design process and shows how to apply this pervasive information architecture methodology to a real-life project with the help of a sample case study. Part 3 contains the chapter *Designing Cross-channel User Experiences,* which brings all the elements of Part 2 together into a single design framework.

A complete and rather eclectic reference section concludes the book.

STRUCTURE OF THE CHAPTERS

All chapters share a common narrative tone; the central chapters which deal with the heuristics (Part 2, chapters 3–8) also share a common structure. They start out with a *short story*, which introduces the central theme, and then move on to tackle the most relevant issues connected to that theme, either by using examples or by referring to literature. Once all the pieces are in place, they position the *theme* in the context of pervasive information architecture. A short recap, in the form of *lessons learned*, follows. This is usually a short bullet list of things we have learned and things we should do when designing. One or more detailed *case studies* cap the chapter. Finally, we wrap it up with a *list of articles, books, and videos* for those who want to track down some of the specific ideas exposed in the text. Chapters in Part 2 and Part 3 also include contributions from a number of international authors, researchers, and practitioners. These usually pinpoint specific concepts or provide examples and case studies that illustrate a relevant application of the theoretical principles enunciated in the text.

NAVIGATING THE BOOK

We are information architects. Even though reading on paper has its rules and a certain degree of sequentiality cannot be avoided, we wanted to be able to give readers some freedom in the way the book can be browsed, which is usually the purpose of the index and table of contents, and books are usually good at this: we just decided to push the envelope a little. You will find that you can actually just jump from image to image and find out that the core ideas from the various chapters are there, easily recognizable. Then, as many other books from Morgan Kaufmann, we have small side boxes that highlight or explain some important concepts we are dealing with in the main text. We tried to give them some pace and turned them into a bookmarking or navigating system that you can use to check out the most important concepts more easily or to find your way through the book faster.

ON BEING LEAN

We think we gave our editors and our publisher some headache while trying to accommodate design talk, a narrative approach, and wayward references to a thousand apparently unrelated facts and notions inside the rather rigid structure of one single book. We jumped the fence into more than one neighbor's backyard and more than once: cognitive psychology, architecture, industrial design, service design, linguistics, game design and theory, interaction design, cinema, art history, economics, library science, and informatics. These all have some stage time in the book, but they serve one single purpose: make you understand where those simple design principles we want to make clear for you come from and how they fit together.

As a result, you will encounter a few names and acronyms throughout the book, but apart from very brief descriptions and explanations to introduce the odd new concept or a particularly obscure reference (and mostly off-text, using side boxes), you will not find much in-depth information in the book about these other practices, fields, and disciplines. You might read the term service design a couple of times, spot an occasional reference to interaction design, or witness a few mentions of user experience here and there, but you will never find an entire chapter dedicated to explaining what they are or what you should do with them.

The reason is simple: we didn't think this was what the book was being written for. We wanted Pervasive Information Architecture to be focused: we wanted you to have a lean, straight-to-the-point guide to the design of these pervasive, cross-channel user experiences in your hands. We wanted to explain why we think information architecture is an important piece of this new holistic view of design. Why thinking of cross-channel user experiences in terms of information spaces is important. Why the shaping of the process is more important than the single interaction. And we wanted this to get through, to seep down.

That meant sacrificing side threads when not strictly on topic and, as a consequence, drop a few interesting opportunities. We still have more than a couple of folders stuffed full with sketches, ideas, links, drawings, and case studies that investigate some of these connections and overlaps. We are those kind of guys who say "you have five seconds, we can explain" and then go on pontificating for hours. It might very well be that you still find that we could have indulged on these walks in the woods a little more. It might be, but it felt wrong, like adding too much water to your coffee or too much milk to your tea. Instead, each chapter will paint the big picture, provide you with the core ideas, and leave everything else to the references you will find waiting at the end of each of them and on the Web.

THE WEB SITE

This book has a companion Web site, Pervasive Information Architecture, which you can visit at http://pervasiveia.com/. We cannot update the book you just bought, but we sure can update the Web site, and we will. A few ideas that didn't make it to the final cut for reasons of pacing and length are already available there, so pay us a visit if you want more of this pervasive stuff. Plus, we will be around, so feel free to drop us a note, ask a question, or leave a comment. We'd love to hear from you.

EXPERIENCING THE BOOK

Andrea's background is in architecture and design; Luca's is in linguistics and semiology. We have been working, researching, and teaching information architecture for more than 10 years. Andrea lives in Sweden, Luca in Italy. Luca is a gourmet and a connoisseur of wines. Andrea not so much, but he can deliver a mean glögg when it's Christmas time. The best thing we can say about this book is that it has thrived on diversity and complicity and on many years of work, research, digressions, dead ends, and reboots. That has surely helped, as complexity is richness, as you will read later on.

The only place where this rule has been broken, where there could be no agreement, no compromise, is in the kitchen. There is no way you can come to terms with making a recipe a democratic, social process, or a menu a collective resolution. When we are at conferences, Luca looks for the quiet places with good food the locals go to, whereas Andrea looks for the places where the action is. We might need a little help from you here.

If you want to experience how Luca has been working his way through the text you are reading, make the book a part of a Saturday evening with friends and make it your main course: it goes well with pasta of Gragnano and some green pepper filet, generously drowned with some Rosso and Sagrantino di Montefalco.

However, if you want to experience how Andrea has been typing away at his desk, make it part of a cold afternoon when it's good to be inside after a long walk: get a comfy chair by the window, a generous cup of black hot coffee, and some pastry. Swedish kanelbullar is a good choice. Light a reading lamp for maximum effect. Then go the Web site, cast your vote, and tell us what the book feels like.

ACKNOWLEDGMENTS

What you have in your hands right now simply wouldn't have been possible without a lot of coffee. Gorging down gallons of hot black tar-like liquid certainly helped. But many, many talented individuals helped us make best use of the lucidity those outrageous and possibly dangerous quantities of caffeine gave us. In no particular order, our deepest, most heartfelt thank you goes to:

Our editors at Morgan Kauffmann Rachel Roumeliotis, David Bevans, and Mary James for the period she was part of the team. Thank you for putting up with us and for being a constant source of good advice.
Our reviewers and draft readers for pointing out where it simply didn't work. We hope we caught everything you thought was unclear, awkward, too verbose, or plain useless. If we didn't, it's our fault. Thank you.
Our contributors: Kars Alfrink, Cennydd Bowles, Stefano Bussolon, Terence Fenn, Chiara Ferrigno and Gianni Bellisario, Claudio Gnoli, Andrew Hinton, Jason Hobbs, Donna Maurer, Eric Reiss, and Samantha Starmer. Thank you for being one of the most amazing international teams ever to grace a book and for providing us with more than one eye-opener.

We also thank Agostino Manduchi, the Tandoori Palace, Noreen Whysel, Jens and Hanna at the library, the Bottoms Up, Davide Potente and Erika Salvini, Benedetta Gizzi, Bruce Springsteen, Sylvain Cottong, Dario Ferracin, Chipmunk, Jacco Nieuwland, Archangel, Carla Campanini, Keith Instone, Stray Cats, Richard Saul Wurman, Badass BBQ, the crew of the Italian IA Summit (Emanuele Dario Alberto Nicola Federico), Monty Python, Melissa Weaver, Umberto Eco, Southside Johnny, Yogi, Andrew Boyd, the Information Architecture Institute, Maria Cristina Lavazza, Tom Waits, Karen Loasby, the Tuschinski, Christian Crumlish, Marcio Bretoni, Raggmunkar, Rat-man, Stephen King, Anders Daniel Jens Mikael Rikard L. Olav Sandra Bertil and

Rikard K., Lou Rosenfeld, the European center for user experience, Gaia Resmini, the Jackie O's Farm, Neal Stephenson, Peter Bogaards, Keith Jarrett and his solo concerts, Alan Moore, GevaliaKid, Jess McMullin, Frank Gehry, Dan Willis, Jorge Arango, Gotan Project, Cristina Trenta, Søren Muus, and the late Willy DeVille.

One special mention goes to Eric Reiss and Jim Kalbach, who in 2006 decided it was worth listening to us and who helped us shape up the idea that would become this book. Thank you.

Finally, thank you big time to Peter Morville for opening up the way and for being a constant source of inspiration and to Dan Klyn for showing us we were on the right track on one hot night in Memphis, Tennessee. Rock on guys. We owe you.

CHAPTER 1

From Multichannel to Cross-channel

FIGURE 1.1
Santa Maria Novella,
Florence.

We are living in an age when changes in communications, storytelling, and information technologies are reshaping almost every aspect of contemporary life—including how we create, consume, learn, and interact with each other. A whole range of new technologies enable consumers to archive, annotate, appropriate, and recirculate media content, and in the process, these technologies have altered the ways that consumers interact with core institutions of government, education, and commerce.

(Jenkins 2005).

SHORT STORY #1: IN 1999

Saturday

It's 1999. Mr. Jones is reading the day's newspaper in the quiet of his apartment in Bridgewater, Somerset, after a light supper. It's an early summer late afternoon on a Saturday, and his wife is in the garden. He is idly browsing the entertainment pages, undecided whether he wants to do some crosswords or not. Something catches his eye: an ad announcing that a documentary about Florence is about to begin in about half an hour on one of the cable channels.

Something on Italy in the Renaissance, by the looks of it, but he and Mrs. Jones have been thinking of taking a week off in Italy for quite some time now. Mr. Jones checks the clock on the wall. Yes, it's 7:18 pm. That's more like 40 minutes then. He passes the news to his wife, reads a little more local news, and when it's just about time he goes to the kitchen to brew some coffee. He carefully measures the coffee. Mr. Jones is 72 and his wife is 69, and they both need to keep it under control when it comes to caffeine in the evening. He sits at the table and waits for the coffee to brew. When it's ready, he pours two cups, puts them on a tray, and brings them to the sitting room. He sits down in his armchair, switches to the right channel, and calls his wife.

The documentary is much better than Mr. Jones thought. Even the coffee is better than he thought. His wife was positively impressed with what they saw and really liked the idea of taking their week off in Florence when he suggested it. A beautiful city, good food, and maybe some tours in the countryside to the gorgeous medieval towns that lie on the hills all around. It's a go, but it's now 9 pm, and Sunday is coming. Mr. Jones will have to go to the travel agency to arrange things on Monday morning, while his wife is at the library where she works as a volunteer since she retired from teaching.

Monday

It's 7:30 am on a sunny and warm Monday morning. Mr. Jones calls the travel agency, but he gets an answering machine that does not tell him what time the agency opens. This is annoying. His wife has already left for the library so he kills some time reading, then he's off. It's a couple of kilometers to the center and to the travel agency, and Mr. Jones is a steady but slow walker. It takes some time, but when he gets there he finds out he still has to wait a little. On Mondays they open late, it seems. Luckily, it's not December. He goes to a café on the other side of the street and gets himself a tea. In some 20 minutes, the agency finally opens.

When he finally sits down in front of the middle-aged woman who runs the place, he finds out things can get a little more complicated than expected. The flight is not a problem, but they will land in Pisa, some 80 kilometers from Florence. That means a local train there, and for some reason, it does not seem possible to buy tickets from England today. The lady reassures him that he will have plenty of time to buy the tickets and that trains run on the hour so that shouldn't be an issue. She also suggests some rather expensive hotel close to the center and the railway station, buffet breakfast included, so they will have everything at hand and staff that can speak English.

Mr. Jones settles for that. After all it's been a while since that trip to Spain in 1995 and the money is not so much an issue, but he asks for a little help in organizing one day out of the city. They check a number of catalogs, but the only package the travel agency can sell him is a complete bus tour of the major medieval cities around Florence that takes 4 days. This is too much for them, as they only have a week and that includes Florence itself. They make a couple of phone calls, but nothing useful comes up. Mr. Jones resolves to look for that once they are in Florence. The agency confirms the tickets and their hotel on Wednesday. Mr. Jones walks back there Friday, pays, and brings all the paperwork home.

Sunday

Their flight lands in Pisa 3 weeks later. It's hot, and they need to find a cab to get to the station to catch the train that will take them to Florence and the hotel. They had started out early to be in London in time for the plane, and they are tired. It's Sunday, and Pisa seems to be rather sleepy. They have the documents and vouchers the agency gave them along with a tiny map of the center of Florence that's not really useful in Pisa. They enter their room almost 4 hours after touching Italian soil, exhausted.

Monday

On their second day, they decide to go to the Uffizi, so just after breakfast they ask the hotel staff for directions. It is pretty close, but they get a little lost in one of the narrower medieval streets; they are not that good with maps. They get there, buy their tickets, queue for an hour, and see their Michelangelo. They dine out. They take pictures at Ponte Vecchio. They buy souvenirs. On their fourth day they even manage to find some sort of shady but actually very nice van tour that takes them to San Gimignano and back. When their week is over and they get home, they have a bag full of tickets, maps, brochures, flyers, and whatnot. They also have five full films to be developed—memories to sort out and share with the Cullings next door. That's what they will do for a few evenings.

SHORT STORY #2: IN 2011

Thursday

It's 2011. It's a late September Thursday afternoon in Trenton, New Jersey, and Mrs. Hutchinson is checking her e-mail. She's in her office and just about ready to leave. She's deleting the usual amount of semispam she receives when she reads one "Check our prices for Florence!" message from a travel Web site she uses for some of her bookings. She and her old high school friend Julie have been talking about Tuscany for a while now, so she checks that out. She finds out there seem to be some good last-minute opportunities for flying to Italy on the weekend. Nothing to blow your mind but enough to make the trip a possibility. She carefully checks the offer and sees that it's either that Friday or never again. She calls Julie on her mobile.

They quickly agree that it can be done if they can find some good central hotel to go with it and if the families can manage an extended weekend without them on such short notice. Mrs. Hutchinson has no children, but Julie has a couple of teenagers in the house and her husband has to agree that he can survive a full 5 days alone with them. They get a green light, and in 20 minutes, after a thorough search through hotel reviews, which gets them a five-star close to the Duomo, which seems good and has a discount rate formula for the weekend, Mrs. Hutchinson is booking the flight and hotel on the travel Web site.

They are landing in Pisa, coming from Munich, Germany, around noon. She checks the location with Google Maps. It's a good 50 miles from Florence. And it's where the Leaning Tower is. It might be worth a stop, if only they had the time. She looks for ways to get to Florence, gets a good deal on a rental car, but does not feel too confident she can drive in the crazy Italian traffic so she leaves that and settles for the train. She buys tickets on the Italian Railways Web site and prints them out carefully. She also prints the timetables. She goes back to Google Maps, sets up a couple of panoramic strolls through the city, and prints these as well. She packs them together with custom maps of all the major places they want to visit, including a couple of restaurants and the Gardens of Boboli and instructions on how to reach them. She then buys tickets for the Uffizi online and calls it a day. Home to prepare her bags.

Saturday

They land in Pisa and arrive in Florence in a couple of hours. They walk to the hotel. They are tired and jet-lagged, but after a couple of hours of sleep and a long shower, they are off for some shopping.

Julie has brought along her digital camera—nothing incredibly professional but enough for them to have a couple thousand pictures from their four days in Tuscany. They will print some and quickly forget about the others. They have a good time.

THE GAME OF THE GOOSE

FIGURE 1.2
A 19th-century game of
the goose board. *Source:
Wikimedia.*

Roughly 10 years separate Mr. and Mrs. Jones's trip to Florence from Mrs. Hutchinson's. Many things have changed in between, even though they all traveled from abroad, visited the city, had a nice afternoon at the Gardens of Boboli, saw Michelangelo's paintings at the Uffizi, and enjoyed some of the countryside.

Mr. Jones had to walk to a travel agency on a working day during its open hours; Mrs. Hutchinson did all her booking on an online travel agency open 24/7.

Mr. Jones had no idea of how to move around or where their hotel was and had to spend some time at the airport looking for a city map and guide in English. Mrs. Hutchinson had printouts of all their movements around the city, and she and her friend Julie spent a couple of hours on the plane to develop some strategies to maximize fun and sightseeing and reduce any unnecessary mileage to a minimum.

Mr. Jones had no control at all over which hotel to choose. He did not have any friendly recommendations and no way to verify what he was offered other than the brochures he was given. Mrs. Hutchinson compared a number of hotels, based on their price, distance, and category. She took a good look at pictures of the hotels, their positions, and their rooms. Some of the pictures were posted by people who spent some nights there. She also received plenty of advice on possible problems (such as asking for proper pillows or more towels) and on tried-and-tested solutions (such as do not go to the desk but rather talk to the maid in charge of the floor).

Mr. Jones brought a few pounds of paper back home, a couple of tourist books, and a hundred pictures. Mrs. Hutchinson brought paper to Florence, used it there, kept a few tickets as souvenirs, bought a couple of ugly miniature replicas of the Palazzo Vecchio, and generally relied on the thousands of pictures that Julie snapped with relentless dedication. They brought home a handful of memory cards.

The 11 years in between the two trips have surely brought an incredible degree of personal control over the details of the journey. If we were to travel to Florence or any other place in the world, we know we could easily compare prices by means of sites such as kayak.com and choose our seats on any plane knowing exactly what the pros and cons are thanks to sites such as seatguru.com. We could see the surroundings of the hotel before booking and read reviews, comments, and tips. We could check for less expensive or more luxurious alternatives without even leaving our chair.

In all, the Internet and the Web have certainly made many activities easier, and this is not limited to traveling, of course: we can shop, make appointments with our doctor, pay our taxes, enroll in higher education courses, and organize events.

FIGURE 1.3

A totally scientific account of Mr. and Mrs. Jones's (top) and Mrs. Hutchinson's (bottom) user experiences in 1999 and 2011.

But have they managed to make all of these experiences more memorable and meaningful or are they still a simple collection of differently shaped building blocks that we can use in a sequence of our own, adjusting our strategies as we go along? We believe the latter is true. Check out the following two sketches (Figure 1.3): they might not be of the highest scientific standard, but they are accurate renditions of Mr. Jones's and Mrs. Hutchinson's respective user journeys.

The various touch points, or interactions with people, objects, or services across the different channels, actually managed to mostly hinder their user experience. For Mr. and Mrs. Jones, at times it felt like they were bouncing off solid walls that had to be climbed. Granted, there is a good deal of difference between the hoops and the loops they had to suffer through and the smoother journey Mrs. Hutchinson and her friend Julie had. The years in between have carved some holes in the walls and have lowered the obstacles. But still, it's a quantitative difference, not really a qualitative difference.

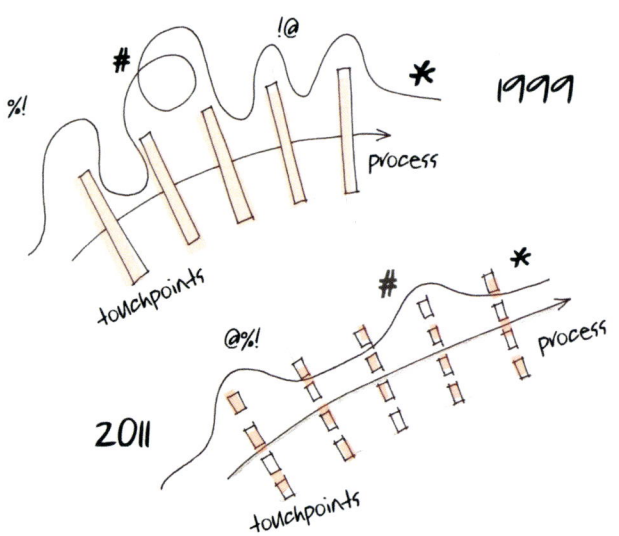

In the face of the technological changes and the incredible increase in available information, arranging a trip like that still feels like we are playing a game of the goose: race from the start to the end and avoid being sent back or missing a turn. It's just that the board is not really a board, but it's channels, media, environments, and experiences and they all have to be played differently. We have to learn a thousand different ways to do the same stuff over and over again, and we cannot play on one single board with one single set of rules. It shouldn't be like this now. Will it be like this in 10 years?

CHALLENGING COMPLEXITY

The unit of analysis for us isn't the building, it's the use of the building through time.

(F. Duffy 1990)

In 2009, MIT researcher Pranav Mistry surprised everyone with the SixthSense wearable interface (Figure 1.4). Composed of a camera, a projector, and a mirror combined into a portable gadget and connected to a mobile computing device that can be pocketed, the open-sourced SixthSense is a veritable piece of design linking digital devices and information with the physical world and making, in turn, as Pranav Mistry puts it, "the entire world your computer."

> Although the miniaturization of computing devices allows us to carry computers in our pockets, keeping us continually connected to the digital world, there is no link between our digital devices and our interactions with the physical world. Information is confined traditionally on paper or digitally on a screen. SixthSense bridges this gap, bringing intangible, digital information out into the tangible world, and allowing us to interact with this information via natural hand gestures.

(Mistry 2009a).

FIGURE 1.4
The SixthSense wearable interface demonstrating augmented reality phone calls. *Source: Pranav Mistry.*

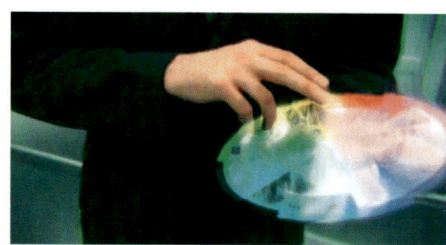

FIGURE 1.5

A paint interface for augmented reality in Bruce Branit's World Builder video. *Source: YouTube.*

It's an amazing appliance, a brilliant glimpse of what we can expect when we start mixing the physical and the digital. We are not that far away from the kind of remediation of reality that Bruce Branit imagined in his World Builder video (Figure 1.5).

Nonetheless, for all its incredible ingenuity, we still believe that there is one larger problem looming behind, one that the SixthSense does not address directly and that deserves our attention first: that of enabling seamless pleasurable, recognizable, and simpler user experiences across channels. The SixthSense brings information into the real world, but it does not address the problem of how that information is designed in the first place. This is a different challenge, one that requires paying attention to a whole different set of design problems.

ACROSS CHANNELS

We have been mentioning the fact that some of Mrs. Hutchinson's experience in Short Story #2 prefigures a cross-channel architecture. What does this mean? In what respect is it different from a traditional multichannel architecture? Well, it is a small but crucial difference.

In traditional multichannel strategies, more than one channel is used simultaneously and alternatively: it's like our friend Mr. Jones being told he could have just phoned the agency and bought a travel package to Florence. The office and phone support are two different alternative channels that can be used in place of each other, at least for certain services. Think of dealing with your bank; you might be able to pay an invoice or file a form by means of several different start-to-end procedures, usually residing in different domains, for example, by calling a phone help service, going to the closest bank branch, or visiting the bank's Web site.

Cross-channel - *Cross-media*, or transmedia, is a term that owes a great deal to the pioneering work on convergence of Henry Jenkins at MIT. It generally refers to linking across different media of branded entertainment and content, such as movies, TV shows, advertising, and games. Cross-media content is distributed and broadcast in such a way that any one single medium offers only fragments of the global experience and actively depends on the others for advancing the narrative. However, the term *cross-channel* has been more widely adopted by the marketing and service design communities for those experiences that span media and environments but are not necessarily connected or limited to the content offered by the entertainment industry.

In **cross-channel**, a single service is spread across multiple channels in such a way that it can be experienced as a whole (if ever) only by polling a number of different environments and media. To keep up with our banking example just given, it's like receiving a text message on your phone giving you the details of some account operation you have performed online and that you need to complete at your bank branch. If one of the pieces is missing, you might miss some of the information being transmitted along the process and that may or may not be available through other channels.

This is where we are moving to. More and more information is reverberated through different channels

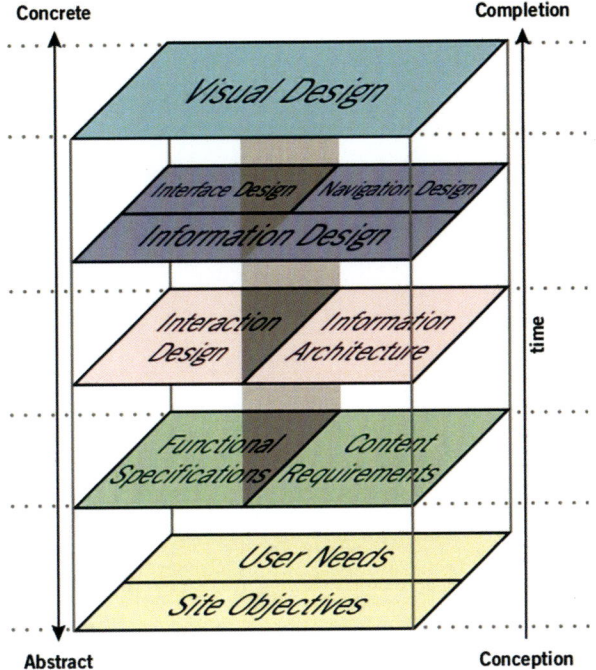

FIGURE 1.6

Jesse James Garrett's original diagram documenting how the Web as a software interface and the Web as hypertext are joined together in a single workflow in user experience. *Source: J. J. Garrett*, The Elements of User Experience, *New Riders Publishing 2002.*

and media: our perception of the process and our expectations of its outcome are changing. We are becoming more aware of its cross-contextuality. In 2002, Jesse James Garrett, a user experience designer and the man who coined the term *AJAX*, wrote a pivotal book called *The Elements of User Experience*. In its pages Jesse explained in detail his model of user-centered design, the one you can see in Figure 1.6.

Jesse identified two parallel forces or areas in the design of user experience that he called *Web as software* and *Web as hyperlink*.[1] These roughly coincide on the one hand with the technological issues, steps, and expertise and with the content-related parts on the other, respectively. Every project moves from conception to completion, developing through time and a series of planes, or activities, that become increasingly more concrete and less abstract as you move toward the final product. Close to the bottom you can find, for example, functional specifications (Web as software) and content requirements (Web as hyperlink). A couple of steps up you find interaction design (Web as software) and information architecture (Web as hyperlink). At the top, visual design completes the picture.

---

[1] While originally aimed at Web sites, this model has been largely applied to user experience in different domains.

A

B

FIGURE 1.7
From single silos to designing across channels.

These are the blueprint for building a silo, and it's ok if you are developing one single artifact. It still works like a charm. But what happens if you have more than one? What happens if your design has to consider more than one media or platform? What happens when your design is cross-channel (Figure 1.7)? Let's go see.

RESOURCES

Articles

Jenkins, H. (2005). Media Convergence. http://web.mit.edu/cms/People/henry3/converge.html.

Mistry, P. (2009a). SixthSense: Integrating information with the real world. http://www.pranavmistry.com/projects/sixthsense/.

MIT. (2008). Convergence Culture Consortium. http://convergenceculture.org/.

Books

Brabazon, T. (Ed.). (2008). *The Revolution Will Not Be Downloaded*. Chandon Publishing.

Briggs, A., & Burke, P. (2002). *A Social History of the Media*. Polity Press.

Garrett, J. J. (2002). *The Elements of User Experience*. New Riders Publishing.

Jenkins, H. (2006). *Convergence Culture: Where Old and New Media Collide*. New York University Press.

Sunstein, C. R. (2008). *Infotopia*. Oxford University Press.

Videos

Branit, B. (2009). *World Builder*. http://www.youtube.com/watch?v=VzFpg271sm8.

Mistry, P. (2009b). *The Thrilling Potential of SixthSense Technology*. http://www.ted.com/talks/pranav_mistry_the_thrilling_potential_of_sixthsense_technology.html.

CHAPTER 2

Toward a Pervasive Information Architecture

FIGURE 2.1
Ranakpur Jain Temple,
India. Photo: M. Savage.
Source: Flickr.

THE ELEPHANT AND THE BLIND MEN

So oft in theologic wars/The disputants, I ween,
Rail on in utter ignorance/Of what each other mean,
And prate about an Elephant/Not one of them has seen

(Saxe 1873).

A well-known traditional Indian story recounts the tale of how a group of blind men made their first acquaintance with an elephant and consequently understood how truth can be an elusive concept. As with all traditional stories, it has variations and local versions, and although at least in the English-speaking world Saxe's poem is possibly the best-known rendition, what we present here is the longer, narrative, Jain version of the story. Jainism is a Dharmic religion, professing nonviolence as a means to elevate one's spirit, and the story of the elephant and the blind men is meant primarily to illustrate how it is possible

Truth can be perceived in different ways

to live in harmony with people who have different beliefs and how **truth can be perceived in different, nonantithetic ways**. Get your popcorn ready; it goes like this:

> Once upon a time, there lived six blind men in a village. One day the villagers told them, "Hey, there is an elephant in the village today." They had no idea what an elephant is. They decided, "Even though we would not be able to see it, let us go and feel it anyway." All of them went where the elephant was. Every one of them touched the elephant.
>
> "Hey, the elephant is a pillar," said the first man who touched his leg.
>
> "Oh, no! it is like a rope," said the second man who touched the tail.
>
> "Oh, no! it is like a thick branch of a tree," said the third man who touched the trunk of the elephant.
>
> "It is like a big hand fan," said the fourth man who touched the ear of the elephant.
>
> "It is like a huge wall," said the fifth man who touched the belly of the elephant.
>
> "It is like a solid pipe," said the sixth man who touched the tusk of the elephant.
>
> They began to argue about the elephant and every one of them insisted that he was right. It looked like they were getting agitated. A wise man was passing by and he saw this. He stopped and asked them, "What is the matter?" They said, "We cannot agree to what the elephant is like." Each one of them told what he thought the elephant was like. The wise man calmly explained to them, "All of you are right. The reason every one of you is telling it differently is because each one of you touched a different part of the elephant. So, actually the elephant has all those features what you all said."
>
> "Oh!" everyone said. There was no more fight. They felt happy that they were all right.
>
> (Jainism Global Resource Center).

The moral of the story is that there may be some truth to what someone says. Sometimes we can see that truth and sometimes not because they may have a different perspective that we may not agree to. So, rather than arguing like the blind men, we should say, "Maybe you have your reasons." This way we don't get in arguments. In Jainism, it is explained that truth can be stated in seven different ways. So, you can see how broad our religion is. It teaches us to be tolerant toward others for their viewpoints. This allows us to live in harmony with people of different thinking. This is known as the Syadvada, Anekantvad, or the theory of Manifold Predictions.

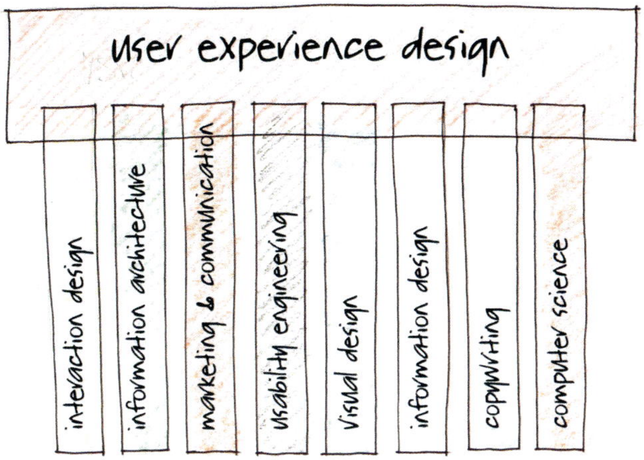

FIGURE 2.2

Boersma's T-model diagram showing vertical specialized silos and the broad overlapping horizontal UXD area.

If you have ever been at any conference or meeting you possibly witnessed discussions like this—and more than once. In the ever-stirring landscape where the fields and disciplines that deal with the design of information-rich services, products, and environments reside, relationships are uneasy, and the story of the elephant is a perfect metaphor readily understood.

In the most accredited view, interaction design, usability engineering, information visualization, content strategy, graphic design, content management, and information architecture are parts of the elephant. As far as we are concerned, we agree: user experience is the elephant.[1] Peter Boersma's famous **T-model** (Figure 2.2) (Boersma 2004) is, after all, a very well-executed retelling of this story for techies and design geeks, and we like the T-model.

Mind you, we are not trying to say that this is an easily settled quarrel: we are just trying to say that it's a different book. The user experience community at large has more brilliant blind men and women groping around than South Africa elephants to hand out for enlightening them. But all field-fencing and name-calling aside, as every large, lively, opinionated community has its share of vocal practitioners who are not afraid to cry wolf now and then, this pachydermic view purports a certain equilibrium and perfectly satisfies our very specific, tedious problem of addressing and identifying fields, areas, and contributions. Our only concern is to **move things** a little farther down the road.

T-model - Peter Boersma, a Dutch user experience designer, developed the idea of the T-model on a restaurant napkin while having dinner with the Information Architecture Institute board member Eric Reiss. It was partially a reaction to Peter Morville's thoughts on Big IA. Boersma thought that graphically representing information architecture as one dominant field was a mistake. He also had the intuition that every field in the area of digital design has its own variant of the Big vs. Little debate. In his diagram, then, all vertical silos (the Little zone) stand back to back and share the overlapping horizontal line (the Big zone). For Boersma, horizontal overlap is the place where user experience design happens.

Moving things farther

---

[1] The most momentous embodiment of this vision is certainly Jesse James Garrett's closing plenary at the ASIS&T 10th Information Architecture Summit in Memphis (Garrett 2009).

Plus, and it's a big plus, we stay clear of a nasty and largely uncharted mine-field, we can live to see another day, and we get to spend a little more of our time trying to illustrate how we went from making one-page Web sites prettier to pervasive information architecture. Not a bad deal.

FROM HUMAN–COMPUTER INTERACTION TO HUMAN–INFORMATION INTERACTION

The design of good houses requires an understanding of both the construction materials and the behavior of real humans.

(Morville 2002b).

We use computers and intelligent appliances primarily to access, produce, and consume information. Computers are tools for the mind that you interact with using your body, or parts of your body. It is not exactly surprising then that "interacting with computers" has historically been some sort of divided king-dom where **information retrieval** and **user interfacing** ruled over two very distinct, noncommunicating hills.[2] The product of a time when computers were limited, relatively slow, and not very interactive, the Kingdom of How You Operate Computers and the Sultanate of Retrieving Information from Computers were autonomous states with different statutes, different customs, and different citizens and subjects. Nonetheless, they shared an underlying assumption that interactions, that is, the use of computers for a given goal, were to happen in absolute, precise propositions to be then conveyed through a stand-alone computer screen.

Information retrieval and user interfacing

The rigorous use of analysis to "break down the whole problem into compo-nents and first focus on the components that promise to yield," which Gary Marchionini – professor at the School of Information and Library Science at the University of North Carolina where he heads the Interaction Design Laboratory – considers one of the key elements of information retrieval,[3] shows initially an intentional fundamental concern with information objects only – which can be pinned down with relative ease – and a certain distance from the people who work with those objects and who are far "less predictable and more dif-ficult and expensive to manipulate experimentally." This has been changing, if slowly: as Marchionini (2004) noted, a new paradigm of information interac-tion has emerged, as

all objects (are) becoming more dynamic and less static and dependable for IR purposes. For example, an active blog is an object

---

[2] This distinction is still clearly reverberating in Jesse James Garrett's model mentioned in Chapter 1.

[3] Information retrieval is often shortened to IR and user interfaces to UI. Information architecture is IA, user experience is UX, information systems is IS, and interaction design is IxD, not to confuse it with industrial design.

that is continually changing and its representations must likewise be continually updated as well. This change emanates from new capabilities within objects and from new capabilities in the external environment that contains them. Internally, electronic objects increasingly are designed to exhibit behavior—to 'act' according to external conditions. . . . Additionally, the system may save increasingly detailed traces of fleeting ephemeral states arising in online transactions—perhaps as extreme as client-side mouse movements as well as clicks. Thus, our objects acquire histories, annotations, and linkages that may strongly influence retrieval and use.

This paradigm was identified in 1995 by Nahum Gershon and was named *Human–Information Interaction* (HII), the study of how human beings interact with, relate to, and process information regardless of the medium that happens to connect the two.

Conversely, on what we could call the traditional design side, a significant shift toward the idea of interfaces as artifacts came about in the field of design only in the late 1980s and early 1990s, when human–computer interaction (HCI) was already a long-standing, acknowledged if specialty area in computer science. So far, design was mostly concerned with chairs, lamps, and buildings: the idea that an **interface could be an object of design**, an artifact, was rather radical. Gui Bonsiepe, a German-born designer, once a teacher at the post-Bauhaus Hochschule für Gestaltung in Ulm, Germany, was among the first to foresee that a design issue was at hand: how it is possible to bring together such heterogeneous parties as the human body, the goal of a given action, and an artifact or a piece of information in the context of communication.

Interfaces as artifacts

His answer was that the binding magic was the interface: not an item per se, but a space in which the interaction among the human body, the tool (the artifact, regardless of its being a factual object or a communication object), and the goal could be expressed (Bonsiepe 1995).

Even so, and even when very lucidly forecasting the possibility, rather contrived at the time, of what we call today *information overload*, having so much information at hand that it's like having none, Bonsiepe was thinking inside the box of industrial design. His views on the impending rise of *infodesign* (as he called it) were somehow limited to the then relatively new field of user interfaces for software programs and to traditional broadcast media: upcoming changes were limited to the absolute frame of a static, mainframe-inherited, and desktop-driven future.

In the late 1990s and early 2000s it became clear that this was not going to be the case for very long: fast, connected microcomputers were first discussed and then used everywhere: inside mobile phones, cars, low-budget cameras, home appliances, ticketing systems, video and music players, you name it. They were capable of communicating with other devices using a range of different technologies—ethernet, USB, wi-fi, GSM, Bluetooth.

They were also becoming actors on the World Wide Web.

Ubiquitous computing - The term refers to embedding computing power into the environment, in objects, appliances, displays, and systems, mostly in invisible ways. The term is universally attributed to Mark Weiser, who in 1991 wrote an article entitled "The Computer for the 21st Century," which was published by *Scientific American*. In ubiquitous computing, users interact with many different devices and systems, in different places, with varying degrees of awareness that they are using computer tools. Pervasive computing, ambient intelligence, smart things, physical computing, and the Internet of Things are all terms for ubiquitous computing used mostly interchangeably but actually emphasizing different points of view—technical, social, or computational—on the phenomenon.

This is what **ubiquitous computing, ambient intelligence,** and **pervasive computing** are all about. As a result, interaction is not limited to precise queries run through specialized interfaces in a controlled environment by engineers in white aprons or to dealing with a few well-known office automation software programs: interactions happen everywhere and reverberate on the Web.

As a consequence, attention started to shift from retrieval, interfaces, and associated fields and practices to the design of interactions and user experience, and new paradigms emerged that cared more for user experience and for social communication by means of networked computers or intelligent appliances: information architecture as we intend it in this book is one of them.

Think of it as of a sort of reverse McLuhan meme: in his book *Understanding Media*, written in 1964, Marshall McLuhan famously wrote that "the medium is the message." What this short sentence meant was that the medium influences how a message is perceived by the audience.

Messages have no medium

Today, something you tweet is maybe read via an RSS feed by a friend using a mobile device with some third-party app you know nothing of and then bounced on to different channels and platforms a dozen or more times. It's an evolving scenario where fast replication and forwarding of pieces of information are the rule: from mobile to desktop, from central to personal, from Twitter to Facebook to FriendFeed to e-mail, and where mash-ups and remediation happen constantly. Here **messages have factually no medium** and the message is once again just the message.

Bridge experience - The term was first introduced by Joel Grossman in an article for *UX Matters* in 2006. A bridge experience is "one in which the user experience spans multiple communications channels, document genres, or media formats for a specific, tactical purpose."

This has three different implications: (1) as messages become messages again, direct human–information interaction becomes a mainstream issue; (2) their being rebroadcast and remediated constantly introduces a shift from traditional multichannel strategies toward **bridge experiences** and cross-channel scenarios; and (3) both human–information interaction and cross-channel design prefigure a process-oriented solution as they necessarily introduce a holistic, global approach.

Working on the general structure across channels allows designers to introduce global, constant cognitive (or interaction) patterns that single artifacts, single touch points in the process, can use and exploit without forcing users to learn or relearn multiple diverging behaviors even when individual interfaces differ. Information architecture plays a crucial role in the design of such systems. But then, what is information architecture?

A BRIEF HISTORY OF INFORMATION ARCHITECTURE

The metaphors we use constantly in our everyday language profoundly influence what we do because they shape our understanding. They help us describe and explore new ideas in terms and concepts found in more familiar domains.

(Morrogh 2003).

Information architecture, or in short IA, is a professional practice and field of studies focused on solving the basic problems of accessing, and using, the vast amounts of information available today. You commonly hear of information architecture in connection with the design of Web sites both large and small, and when wireframes, labels, and taxonomies are discussed. As it is today, it is mainly a production activity, a craft, and it relies on an inductive process and a set, or many sets, of guidelines, best practices, and personal and professional expertise. In other words, information architecture is arguably not a science but an **applied art**. Very much like industrial design, say.

Information architecture as an applied art
1964: IBM

Even though its modern use, strictly related to the design of information, goes back no farther than the mid-1970s and **Richard Saul Wurman's famous address** at the American Institute of Architecture conference of 1976, use of the term *information* together with the term *architecture*, has been around for a little bit longer and in quite a few different settings.

In an **IBM research paper** written in 1964, some 12 years before Wurman, and entitled "Architecture of the IBM System/360" (Amdahl et al. 1964), *architecture* is defined as

the conceptual structure and functional behavior, distinguishing the organization of data flows and controls, logical design, and physical implementation.

Wurman at the AIA - Wurman wrote an article with Joel Katz entitled "Beyond Graphics: The Architecture of Information," which was published by the *AIA Journal* in 1975. In an interview with Dirk Knemeyer in 2004, Wurman said: "The common term then was 'information design.' What got confusing was information design and interior design and industrial design, at that moment and still today in many and most people's minds, are about making something look good. Interior designers make your place look better, industrial designers were engineers doing something that usually went to an engineer to put a package around it. Information design was epitomized by which map looked the best—not which took care of a lot of parallel systemic parts. That is what I thought 'architecture' did and was a clearer word that had to do with systems that worked and performed.... I thought the explosion of data needed an architecture, needed a series of systems, needed systemic design, a series of performance criteria to measure it. There are thousands of people using the term [*information architecture*], and they have no idea where the term came from, and 90 percent of them aren't doing what I think they should be doing anyway."

It is not disputable that we are talking computer architectures here, disks and boxes and wires and hubs, but the way in which the term *architecture* is abstracted and conceptualized in connection with *structure* and *behavior* and not just physical layouts laid the basis for the subsequent extension of its use to other areas of computing.[4]

A few years later, in 1970, at the Xerox Palo Alto Research Center (PARC), a group of people specialized in information science was assembled and then given the charter to develop technology that could support the "**architecture of information**" (Pake 1985). This group was single-handedly responsible for quite a lot of important contributions, including the first personal computer with a user-friendly interface, laser printing, and the first WYSIWYG text editor. As Marti Hearst, now a professor at the University of California Berkeley, recalls,

1970: PARC and the architecture of information

perhaps because of the social nature of information creation and use, much of the technical research at PARC has emphasized the human-computer interaction and social aspects of computing.

Louis Murray Weitzman (1995), in his Ph.D. research at the Massachusetts Institute of Technology on "The Architecture of Information: Interpretation and Presentation of Information in Dynamic Environments," supports this notion that the modern conception of the term originally came from Xerox Labs.[5] Quoting Smith and Alexander's *Fumbling the Future* (1988), he reports that

Xerox was among the first corporations to address this notion of information structure and use the "elegant and inspiring phraseology, the architecture of information" to define its new corporate mission.

This high-level framing, the necessity for a broader vision, remained one of the core concepts for the many who wrote about and discussed the development of information architecture up to the mid-1980s. It might seem a small thing, but Cuban information architect Rodrigo Ronda León (2008) correctly demonstrates how this seminal documentary evidence of the use of the terms *architecture* and *information* together joined straight from the start specialists in information science and in user-focused development, a trait that would be somehow brought on to greater visibility and results only in the 1990s.

Focus on the user and information science together from the very start

---

[4] Much of this discussion owes a great deal to the work of Rodrigo Ronda León (2008).

[5] In addition to providing further documentary evidence to support this notion, Weitzman also underlines how Xerox actually contributed vastly to the general view of information architecture as a tool to support the design and presentation of documents, something that is of vital importance in Wurman's work.

After Wurman and his stunt at the American Institute of Architects (AIA), though, information architecture seemingly went through a dormant stretch all through the 1980s, a period during which the idea of information architecture as the design of complex or dynamically changing information seemed to be lost to a view much more akin to that of **information systems**. Articles written in the 1980s mostly refer to information architecture as a tool for the design and creation of computer infrastructures and data layers, with a larger emphasis on the organizational and business aspects of the information networks (Morrogh 2003). Curiously enough, most of the underlying nuts and bolts associated with information architecture design today are actually a product of this period: blueprints, requirements, information categories, guidelines on the underlying business processes, and global corporate needs—they all made their way into IA-related territory in the 1980s (Brancheau & Wetherbe 1986). They were incorporated once and for all in the information architecture toolkit by the wave of the late 1990s led by Rosenfeld and Morville.

1980s: Information systems

From these and other observations, Ronda León derives a graphical chronology of IA events, mainly books, papers, and conferences, and a three-part development hypothesis (Figure 2.3) spanning roughly 30 years, in which the two early phases, that of information design (1960s–1970s) and that of system (and systemic) design (1980s), are integrated into the modern mainstream idea of information architecture as we know it today.

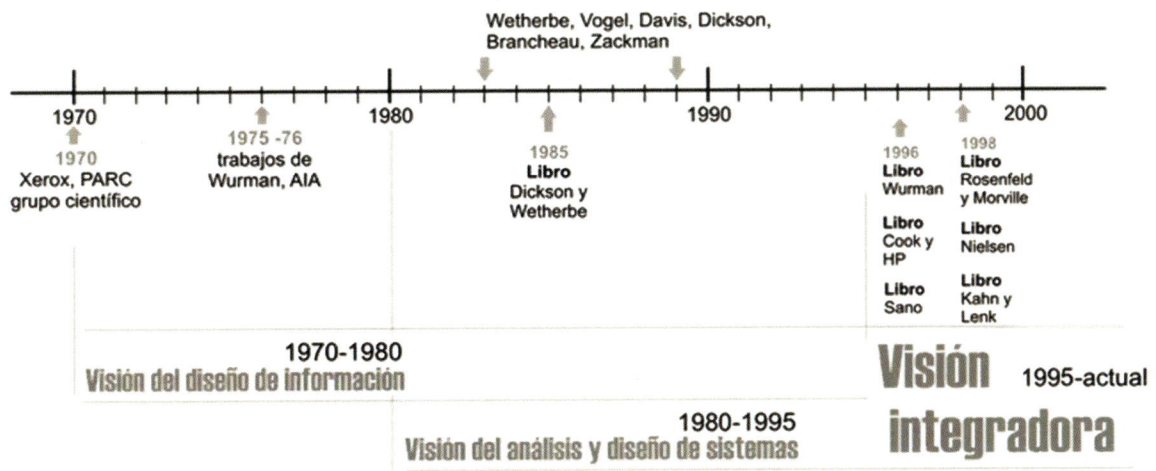

FIGURE 2.3

A chronology of information architecture in the 1980s and early 1990s. *Source: Ronda León (2008).*

We will return to this timeline toward the end of the chapter.

Whatever you might think of it, it is clear that the early take on information architecture that developed from the IBM paper, PARC, and Wurman's initial vision had to wait until some missing pieces entered the scene. One of those pieces was certainly the possibility for professionals to **operate on large amounts of data** in a new media, void of or minimally encumbered by preexisting corporate hierarchies.

1990s: The Web as a catalyst

The World Wide Web provided just that one-time chance. That is why worldwide acknowledgment that something called *information architecture* was there to help redefine first Web design and then the way we design human–information interaction came in 1998 when Louis Rosenfeld and Peter Morville, both of them librarians, wrote *Information Architecture for the World Wide Web*, affectionately called the Polar Bear book because of the polar bear on its cover and one of the cornerstone books in the field.

APPROACHES TO INFORMATION ARCHITECTURE

That's why I've chosen to call myself an information architect. I don't mean a bricks and mortar architect. I mean architect as used in the words architect of foreign policy. I mean architect as in the creating of systemic, structural, and orderly principles to make something work— the thoughtful making of artifact, idea, or policy that informs because it is clear. I use the word information in its truest sense. Because most of the word information contains the word inform, I call things information only if they inform me, not if they are just collections of data, of stuff.

(Wurman 1997).

Information architecture as a way to tackle new and unforeseen issues

Given this long coming together, it seems just ordinary that information architecture has its roots in a rather large number of different disciplines: information design, visual design, library and information science, cognitive psychology, architecture, and probably a few others. As a rule, specialized fields emerge out of the necessity to dig deeper and in specific directions inside an already established field: for example, environmental psychology was born to better investigate, in an interdisciplinary setting, the psychological interplay between human beings and their surroundings and could rely on a preexisting framework of theories and ideas. This was not the case with information architecture, which emerged as a way to tackle issues that were, or seemed to be, totally new, unforeseen, and **requiring pioneer thinking**. As Andrew Hinton (2008)—Information Architecture Institute founder and one of the best thinkers in the information architecture community—argues in *Linkosophy*, we have been having hypertext for quite a long time alright. But worldwide, readily available networked hyperlinking is a different thing.

This book is not a comprehensive critical effort or an effective analytical assessment of the cultural, theoretical, and methodological history of information architecture: we won't even get close to the thing if it was lying by the side of the road with its belly up. We know it's just playing dead and that it has a vicious bite. But is it possible to draw some lines? Figure out some patterns?

Do not worry: when we said that this is a book on design, on doing stuff, we did not lie. But trying to understand, at least in generic terms, from whence we came can certainly help us see where we are going. Plus, we have no need to complicate things more than necessary: we stick to Ronda León's chronological structure, we have a basic scheme in place that we do not even need to twist and bend too much. The three consecutive periods in the timeline can effectively translate to the three broad, different approaches that have characterized information architecture so far, with the differentiating factor being the way they work with the core resource of the field, information. Because we are resourceful and imaginative chaps, we decided to call these the information design approach, the information systems approach, and the information science approach. Let's take a look at them in detail.

Information Design

This roughly corresponds to Richard Saul Wurman's contribution and initial vision. For Wurman, design and architecture are the basis for a science and art of creating "instruction(s) for organized space" (Wurman 1997) and for making these **understandable**. Wurman published his seminal book *Information Architects* in 1997, just one year before Rosenfeld and Morville's IA bible hit the shelves: the book dealt with the increasing difficulty Wurman was experiencing in communicating the rising amounts of information, Wurman's style, and presented a large selection of design solutions to the problem. It was a designer's book: from a designer, for designers. As much as architects are expected to create structure and order in the world through planning and building, information architects were expected to draw lines and derive some kind of order in data space for better understanding and enjoyment, their primary task being to make this information simpler, more direct, and more comprehensible.

Information architecture as understanding

According to what he said to Dan Klyn of University of Michigan in a recent set of interviews, Wurman had no master plan in mind when he rolled out information architecture at the national conference of the AIA: he was just trying to "find patterns for himself."[6] Neither was he interested in disseminating his ideas to a new audience or in creating a new field or profession, and he was

---

[6] The interview contains this brief passage: (Klyn) "Did you intend to create a movement within the field of architecture to focus on information display and organization and such things?" (Wurman) "No."

actually quite surprised (not to say a little upset) when he finally did find out what an echo his ideas and activities had managed to stir up.

At the time, Wurman gave an extremely precise definition of *information architect*, which still largely holds up today:

> a. the individual who organizes the patterns inherent[7] in data, making the complex clear; b. a person who creates the structure or map of information which allows others to find their personal paths to knowledge; c. the emerging 21st century professional occupation addressing the needs of the age focused upon clarity, human understanding, and the science of the organization of information.

Static design of large quantities of information

Even though he was possibly mainly concerned with the **static design** of large quantities of visual information, his contribution was undoubtedly a major source of inspiration in the initial modern reframing of the field (Wodtke 2002) when it was moved to the Web.

Wurman finally came to terms with his being considered part of the ongoing information architecture conversation in 2010 when he was invited to keynote at the 11th ASIS&T IA Summit in Phoenix, Arizona (Figure 2.4). He definitely did not disappoint the audience (Wurman 2010).

FIGURE 2.4
Wurman showing how to peel a banana on stage at the 11th ASIS&T Information Architecture Summit in Phoenix, Arizona.

---

7 *Inherent* is still the one single word we do not like in this definition. Andrea has written a few thoughts on this in a post entitled "Of Patterns and Structures" that you can read on his blog. See References.

Information Systems

This approach is tightly connected to the developments of the 1980s and the logic of what has become today information systems and business informatics much more than to the logic of user experience, and it concerns itself mostly with ways to connect problems of **information management** within the larger business vision or logistic needs that drive organizations.

The widely recognized semantic shift toward user experience that followed the Polar Bear book has made "information systems information architecture" a minority but important stance, which still manages to be prominent in large corporate settings, fiercely attached to their IT roots, and consequently produces quite some friction whenever it gets in the same room with mainstream information architecture.

Gene Leganza's report on information architecture published for Forrester Research in 2010 represents these views well. The 20-odd-page document clearly defines how the information architect role is primarily an IT function, as its main task is to enable consistent access to correct data and splits information architecture into two very distinct concepts: one is the "structuring of all enterprise-wide information assets," which is *enterprise IA*, and the other is the design of "information for an individual Web site, portal, or application UI," which is "user experience IA," or "Web IA."

Leganza also states that the value of IA is still not self-evident to many an enterprise architect (with 43% of them not really considering the domain part of their strategies) and that "the value in IA's structuring the information in an enterprise is not in attaining some abstract goal of imposing order on disarray but in enabling the provisioning of the right information in the appropriate context to the stakeholders who need it," which, by the way, is not really that different from anything you are going to read in this book and in many others that deal with *UX information architecture.*

This enterprise-layered view is not just Forrester's: whenever you deal with company assets, information architecture comes out as

> a holistic way of planning which meets the organization's information needs and avoids duplication, dispersion, and consolidation issues. The information architecture is the collective term used to describe the various components of the overall information infrastructure which take the business model and the component business processes and deliver information systems that support and deliver it. Prime components are the data architecture, the systems architecture and the computer architecture.
>
> (Carter 1999).

In a way, it's a one step up and one step down on the ladder kind of view. It's up, as it connects information architecture to the strategic company thinking that is behind the idea of enterprise or enterprise-level information architecture in a way that "UX IA" has not yet managed to do. It's down, as it actually moves design thinking quickly toward questions of data connections, bandwidth, costs, server topology, and storage limits that are not normally part of the mindset of IAs and that tend to be rather specific and technological in nature.

Information as a resource

We have no particular qualms with this view, but would like to point out that it seems to stem from a totally different assumption that we are not sure we totally agree with. While both the information design and the information science approaches see information as raw material used to build artifacts, the **information systems approach does not**. For those using this approach, as Roger and Elaine Evernden (2003) wrote in their book *Information First*, information architecture is

> a foundation discipline describing the theory, principles, guidelines, standards, conventions and factors for managing information as a resource.

This is possibly the single best way to explain that different angle: it is not much about the design of information per se, but it is about everything that helps manage information as a resource for enterprise consumption and use.

Information Science

This is where Rosenfeld and Morville initially came from. In an interview with Scott Hill (2000) for O'Reilly, they stated that

> in 1994, before the Web took the world by storm, we were teaching some of the first academic and commercial courses about the Internet. We both believed the Internet would become an important medium and that librarians had a great deal to offer this brave new world of networked information environments.

At the time dot-coms were booming, everyone wanted to be on the Web and get rich, and their firm, Argus Associates, originally owned by Rosenfeld and Joseph Janes with Morville an employee, was already a significant player in the U.S. market (Figure 2.5).

FIGURE 2.5
Lou Rosenfeld (left) and Peter Morville (right) in 2000, with Samantha Bailey, then vice president of consulting operations at Argus Associates. Photo: P. Morville.

Rosenfeld and Morville came to use the label *information architecture* from a totally naïve perspective: they were not familiar with Wurman's work at all. In the words of Morville (2004), they

> found (them)selves using the architecture metaphor with clients to highlight the importance of structure and organization in website design. Lou got a gig writing the Web Architect column for *Web Review* magazine, and I soon joined in. In 1996, a book titled *Information Architects* appeared in our offices. We learned that a fellow by the name of Richard Saul Wurman had coined the expression "information architect" in 1975. After reading his book, I remember thinking "this is not information architecture, this is information design."

This is an accurate and insightful statement. Their initial view was entirely focused on the new dynamic environment of the World Wide **Web**, and it certainly had little in common with the *information design* approach. Organization, labeling, navigation, and search were the touch points around which they structured their practice. In that interview with Scott Hill, Rosenfeld said explicitly that these were the key concepts to address in order to

Information architecture for the Web

> help people find and manage information more successfully. Organization systems are the ways content can be grouped. Labeling systems are essentially what you call those content groups. Navigation systems, like navigation bars and site maps, help you move around and browse through the content. Searching systems help you formulate queries that can be matched with relevant documents.
>
> (Hill 2000).

Very famously, they remarked a few years later that the real difference they could see between their view and Wurman's, post hoc, was that for them information architecture was very much the design of what was *between* the pages of a Web site, meaning the links, the structure, and the connections, while for Wurman it seemed to be the design of the pages *themselves*. It could also be said that Rosenfeld and Morville designed for a world of ever-changing content,[8] a somewhat alien idea in Wurman's vision. Not at all surprising, if we consider that at the time Wurman addressed the AIA and started thinking about this information architecture thing Sir Tim Berners Lee was pursuing a degree in physics at the Queen's College in Oxford, England, and getting busy building his first computer "with a soldering iron, TTL gates, an M6800 processor and an old television."[9]

Rosenfeld and Morville, and those many following along their initial LIS view, must be credited for bringing in many of the **core methodologies** used

Completing the vision

---

[8] Or, in perspective, that was becoming dynamic. This will be expanded at the end of the chapter.
[9] As we can read in his longer online biography at the W3C—http://www.w3.org/People/Berners-Lee/Longer.html.

Big IA vs. Little IA - As much as a usually lively but seemingly ordinate general discussion on identity, methodologies, and focus was unfolding in blogs, mailing lists, and conferences, scope became almost immediately a somewhat contentious issue for the growing IA community in the early 2000s. Morville clearly articulated the diverging views in an article entitled "Big Architect, Little Architect" that he wrote for the column Strange Connections and that was published on the Argus Web site in 2000. In this short piece, Morville stated that while there is certainly a core practice of IA that simply involves structuring and organizing information systems for intuitive access to content, the

interpretations of the role of the information architect vary depending upon the organizations, the projects, and the people involved. At one end of the spectrum, the Little Information Architect may focus solely on bottom-up tasks such as the definition of metadata fields and controlled vocabularies. At the other end, the Big Information Architect may play the role of "an orchestra conductor or film director, conceiving a vision and moving the team forward."

This two-step vision was to prove to have the most controversial and gravid of consequences, ruffling quite some feathers, stirring an endless debate with bordering practices, and spawning a thousand discussions, debates, and rebuttals, and finally culminating in Adaptive Path Jesse James Garrett's Memphis Plenary at the 10th ASIS&T IA Summit, where Garrett called for everyone to abandon old useless labels and models and rejoice under the simpler, more appropriate user experience flag, which of course spurred new, slightly reframed debates.

Defining the damn thing

for the design of navigation, labeling, and site structure. They offered the blooming community of practice an extremely empirical and practical approach, and they single-handedly brought user research and usability engineering into the core of mainstream IA tools. As Rosenfeld is fond of saying, they "certainly embraced other disciplines." Through the years their view on the subject evolved, but so far, Rosenfeld and Morville's classical view of IA as the design of taxonomies, menus, and site maps still represents the mainstream and most accredited view, especially for those outside the field.

A DEFINITION OF INFORMATION ARCHITECTURE

The common thought is that architects build buildings. No, architects make instructions for having someone else build them. So basically architects, if you're talking about architects . . . they give instructions.

(Klyn 2009).

As with many other maturing fields, the larger information architecture-aware community, which does not include only information architects, of course, seems to spend a lot of time trying to frame a perfect, static definition of the field itself to the point of having a terribly successful acronym to describe the process: DTDT, for "**defining the damn thing**." DTDT has become both a never-failing in-joke and an incredibly effective party spoiler, capable of generating endless mots d'esprit on mailing lists and Web sites and of making spirited conversations wither and die on the spot.

In an article for the *Bulletin of the American Society for Information Science and Technology*, written when the *Journal of Information Architecture* (Figure 2.6) was about to be launched and entitled "IA Growing Roots," one of the authors of this book suggested that

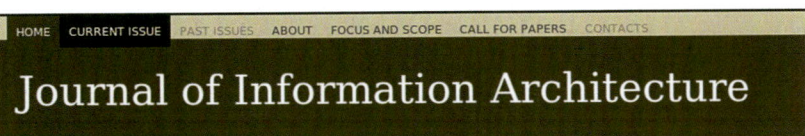

Spring 2009
Issue 1, Volume 1 of the Journal of IA

From Andrew Hinton's The Machineries of Context

> *As we began digitizing our information sources and adding hyperlinks, the information slipped the bounds of physical constraint and started reassembling itself into other structures, many all at once. Still, our minds try to make spatial sense of it, and use spatial memory to organize and keep track of it all, interchangeably making use of semantic relevance and spatial positioning to process our contextual experience.*

> *The web complicates these experiences even further, because its open hyperlinking allows almost any structure imaginable to emerge, confusing the boundaries between the link and the linked. A link to a place becomes part of that place's actual substance. Every link either creates a new context or adds dimension to an existing context. On the web, the map doesn't just make the territory meaningful, the map makes the territory.*

Read Issue 1, Volume 1 of the Journal of Information Architecture »

What is JofIA

The Journal of Information Architecture is an international peer-reviewed scholarly journal. Its aim is to facilitate the systematic development of the scientific body of knowledge in the field of information architecture.

The Journal of Information Architecture is published biannually in English and Volume 1, Issue 1 is the current issue.

Read more about the Journal »

Call for Papers

The Call for Papers for Volume 1, Issue 2, to be published Autumn 2009, is now open.

Read the Call for Papers »

Cover image

A remix of Georgian Dance by flyergeorge, released under the provisions of the CC BY SA 2.0 license.

The Journal of Information Architecture is an independent initiative of REG-iA, the Research & Education Group in IA. It is sponsored by the Information Architecture Institute and by Copenhagen Business School.

REG-iA
http://reg-ia.org/
Information Architecture Institute
http://www.iainstitute.org/
Copenhagen Business School
http://uk.cbs.dk/

Journal of Information Architecture
http://journalofia.org/
ISSN 1903-7260

General Inquiries
info@journalofia.org

Manuscripts and submissions
papers@journalofia.org

Editors
editors@journalofia.org

FIGURE 2.6
The first issue of the *Journal of Information Architecture.*

this "define craze" that regularly seems to seize IAs is somewhat a sign of the times and actually fairly common, and it's a consequence of two different conditions, one internal and the other external: (first), the community is young and somewhat necessarily shallow, and (second) we live in very fast times.

(Resmini, Byström, & Madsen 2009).

Discussing how one's work and research interest should actually be described to both peers and outsiders is far from being an uncommon concern in new communities, but for some reason this is a conversation that makes so many IA pulses race that it is now bordering on hopelessness. The debate on what information architecture is and how it should be defined properly is almost 20 years old and is beginning to rival the one still enveloping "information science" and dating back to the mid-1950s. This is not necessarily a bad thing, though: in an e-mail conversation in the autumn of 2009 (and in many subsequent public occasions), Andrew Hinton pointed out that the "only communities which do not quarrel over such things are dead communities," which is an altogether considerate point.

Definitions help communicate

The fact is, a fair number of those who shun the very topic and drop out of discussion as soon as the dreaded DTDT emerges acknowledge that a definition is somewhat inescapable if anything has to be **communicated**. For them, it's just that opinions on how to reach a common understanding and what this understanding is or should be have been maddeningly oscillating through the years. Adding even more to the confusion, not everyone in the group of those who do not go berserk if you utter a feeble "now, what is IA" agrees that information architecture needs or ever needed a definition. In his opening keynote at the 1st European IA Summit (EuroIA) in 2005, Andrew Dillon, dean of the School of Information at the University of Texas and one of the first academics to take a scientific interest in IA, simply stated that "we don't have a definition for IA and we don't need one."

An art-and-craft approach does not guarantee duplicability

This is a respectable position and is usually connected to a view of the practice (and research) of IA rooted firmly in the profession and in the information architect as a maker, a crafter. Dillon (2005) also recognized the inherent problem with this art-and-craft approach—the **incapability to guarantee consistent reproductions**. Doers often cannot articulate their recipes very well:

> it wouldn't matter what Picasso would say about his process: we still wouldn't be able to paint a Picasso painting. Indeed, crafting is not about the execution of sterile, calculated steps, but rather an intuitive response to a problem.

It's certainly true. It's indeed true of all design disciplines. But as higher education courses and programs that teach or do research into information architecture are being implemented all over the world, students find themselves in need of being taught the essence of the craft, the basic theoretical frameworks, the boundaries of the field, and its role in the creation or cocreation of shared information spaces—in other words, how Picasso painted his paintings. While of course solving this particular do/know conundrum has little or no

JASON HOBBS AND TERENCE FENN—TEACHING THE DESIGN THINKING OF INFORMATION ARCHITECTURE

As teachers, we want our students to learn the thinking skills and patterns needed to solve a broad range of social and commercial problems: wicked problems that can exist in any media, environment, technology, and geography. So how do we unpack *the thinking* involved in designing and how do we teach it?

Increasingly we are finding the skills to be found employed by information architects (for the Web) being applied to these wicked problems. In the theory of design thinking, "placements" are tools by which designers intuitively or deliberately shape a design situation or problem, identifying the views of all participants, the issues that concern them, and the inventions that will serve as working hypotheses for exploration and development.

If you are an information architect, the tools referred to here will resonate with you at a more or less intuitive level. In solving complex informational and interactive challenges on the Web, IAs consider the organization and structure of information in extraordinarily liberal ways, from multiple points of views, with multilayered concerns.

If we decouple the application of information architecture design skills from their application to the Web and information-driven environments, we find ways of thinking that assist in shifting the placement of problems and the tools for their exploration.

The reason for this may reside in the cognitive problem-solving abilities inherent in the act of information architecture design being applicable to solving complex problems found across a multitude of environments. These include, for example, relational thinking, "middle-brain" thinking, categorization, wordplay and semantics, storytelling, structured thinking, and argumentation.

Through a process of teaching, testing, and publishing at the University of Johannesburg, we attempt to discover, explore, and document the latent cognitive problem-solving methods of information architecture in the context of the theory behind design thinking with the aim of developing teaching methods and skills development focused specifically on their use for any media, environment, or combination of environments.

Our hope is that this approach to teaching will facilitate the development of learners who think like information architects rather than simulate the thinking process by an overreliance on fashionable methodologies and trends.

Jason Hobbs runs jh-01/Human Experience Design, a design consultancy in Johannesburg, South Africa. Over the past 13 years his work in user experience design has spanned commercial, nonprofit, arts and culture, and civic projects. He frequently presents at international conferences, is a published author on UX design, mentors, lectures, and works actively to grow the local community of practice in South Africa through the SA UX Forum. Jason is an affiliated researcher at the University of Johannesburg's Research Centre Visual Identities in Art and Design. His Web site is http://www.jh-01.com, and he can be found on Facebook at http://www.facebook.com/#!/profile.php?id=601160572.

Originally fine-art trained, Terence defected to the dark art of visual design in order to gain a scholarship for a master's degree program at the University of New South Wales in Sydney, Australia, in 2002. On his return to South Africa, Terence joined the Department of Multimedia at the University of Johannesburg where he has developed and coordinated the interactive design component of the program since 2003. Primary academic interests include design cognition, structured thinking, and urban typologies. Outside of the Ivory Tower, he enjoys sport, illustration, and traveling. Terence lives with his girlfriend, Jana, and their two dogs in a refurbished old jail in the inner suburbs of Johannesburg.

immediate impact on the professionals and practitioners who are in the market now, it is central for the growth of the discipline and for the professionals of tomorrow.[10]

Others have tried to come up with ways to say what information architecture is by upping the ante and moving into more abstract territories. For example, Earl Morrogh (2003)—designer, writer, photographer, and now a teacher at the College Center for Library Automation in Florida—writes in his book, *Information Architecture, an Emerging 21st Century Profession*, that information architecture is "primarily about the design of information environments and the management of an information environment design process," which again is quite a respectable if slightly fuzzy vision.

The IAI's three-part definition of information architecture

Regardless of this long story of squabbling, arguing, agreeing, and pointing out differences and disagreements that makes information architecture interesting for the not faint of heart, a certain general consensus has been reached over the years on a three-part—initially a four-part—**definition** pushed forth by the second edition of *Information Architecture for the World Wide Web* and officially adjusted and adopted by the Information Architecture Institute (IAI), an international professional organization dedicated to advancing the state of information architecture through research, education, advocacy, and community service, in 2005. According to this canon, information architecture is:

1. The structural design of shared information environments
2. The art and science of organizing and labeling Web sites, intranets, online communities, and software to support usability and findability
3. An emerging community of practice focused on bringing principles of design and architecture to the digital landscape

While the second definition seems to bind information architecture to the confines of World Wide Web–related design, the first and third definitions push the envelope far beyond that. These definitions have largely been commented, dissected, and sometimes openly criticized, but we agree with their basic tenet: that IA as both a practice and a future discipline has more to it than the simple art of labeling and organizing of online content and that it is evolving. Coincidentally, that's what this book is about—evolving information architecture.

---

[10] This is one of the reasons why the authors of this book were both involved in the creation of the *Journal of Information Architecture*, an open access peer-reviewed scientific journal whose main goal is to facilitate systematic development of the scientific body of knowledge in the field of information architecture. The journal is available at http://journalofia.org/.

PERVASIVE INFORMATION ARCHITECTURE

Instability is what fuels the process.

(Soddu & Colabella 1992).

Rosenfeld and Morville's Polar Bear had an enormous success, and in the late 1990s and early 2000s the practice of information architecture was usually synonymous with designing Web sites for the World Wide Web. As 2000 became 2005, things were changing again. Users were entering the scene as producers (or prosumers, as they both consume and produce information), tagging was all the rage, and personal and home devices were starting to redraw the boundaries of what computing was.

Even though a persistent thread kept it tied to the creation of Web-only content, which was (and is) especially true if you move into LIS-connected IA research and practice, a few individuals, people such as Adam Greenfield and Peter Morville, for example, started to consider that this was a limitation with little rationale behind it.

Users were becoming producers, devices were on the move, and **new problems needed to be addressed**. Information architecture was moving into uncharted territories, becoming a boundary practice whose contributions were crucial where complexity, unfamiliarity, and information overload stood in the way of the user, regardless of the very nature of the environment being designed. Information architecture was moving beyond the confines of the Web.

2000s: Information architecture moves beyond the Web

Now have a look at Ronda León's timeline in Figure 2.3 again. Initial view, development, synthesis. We are past that. We could call this three synergistic moment **classical information architecture** and rework that diagram a little (Figure 2.7).

Classical information architecture

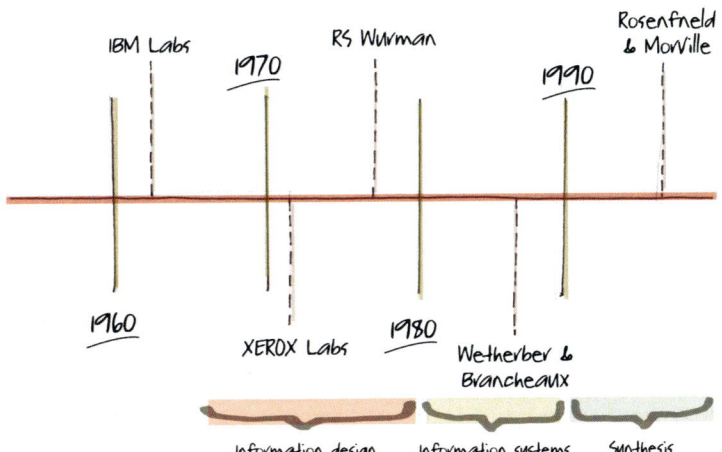

FIGURE 2.7
A timeline for classical information architecture.

Service design - The activity of planning and organizing people, infrastructure, communication, and material components of a service in order to improve its quality, the interaction between service provider and customers, and the customer's experience (from Wikipedia). Sylvain Cottong (@sly), innovator, design thinker, and one of the founders of the European center for user experience (ecux .org), maintains that service design is a dynamic interdisciplinary field joining design, management, and social sciences to provide customers with useful, usable, desirable, attractive, and credible services that get their job done and deliver an outstanding experience without sacrificing feasibility, effectiveness, efficiency, and value from the producer's point of view. Essential to service design is a 360 degree view on touch points and channels where consumers and producers interact.

But what comes after that? Well, a new stage, a new phase, where information architecture becomes pervasive and starts to address the design of information spaces as a process, opening up a conversation with ubiquitous computing and **service design**, and where the information architect recognizes gathering, organizing, and presenting information as tasks analogous to those an architect faces in designing a building (Figure 2.8), as both "design spaces for human beings to live, work, and play in,"[11] with the primary differences being the raw materials they work with (Wodtke 2002). If the architect has to

ascertain those needs (i.e., must gather information about the needs); organize the needs into a coherent pattern that clarifies their nature and interactions; and design a building that will—by means of its rooms, fixtures, machines, and layout, i.e., flow of people and materials— meet the occupants' needs

(Wurman 1997).

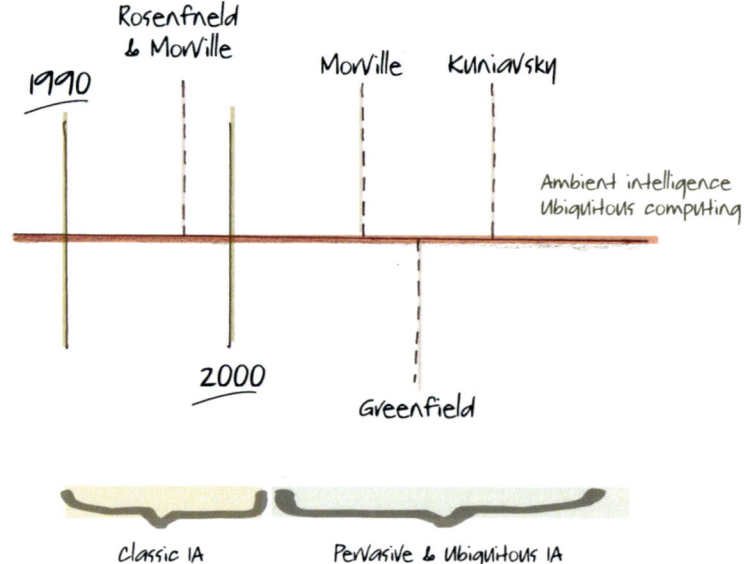

FIGURE 2.8

Moving into pervasive information architecture.

---

[11] With interesting repercussions as well. See Kolson Hurley (2010).

then the information architect has a definitely similar goal in information space, as presenting information for a purpose *is* an architectural task. We design places where people spend an awful lot of time doing stuff, after all, so they better be good.

It might seem an abstract framework, process thinking and a pinch of architectural reasoning, but the bar from what could be thought to be appropriate of simple hypertext systems in the late 1990s has been raised significantly.

Leave the vast implications of the social Web going pervasive behind and consider mobile computing alone: being able to be plugged in on the go means that there is no certainty of the physical context in which a certain piece of information is consumed, which, in turn, is one huge design challenge. The way we interact, the data we need, how we allow ourself to be distracted by the information we receive, and the urgency or timing of warnings or reminders change all the time. But they are always there: we do not simply switch the computer off and walk out.

When we increasingly experience the world through one or many disembodied selves (Inalhan & Finch 2004) and live in a world where relationships with people, places, objects, and companies are shaped by semantics and not (mostly, or only) by physical proximity; where our digital identities become persistent even when we are not sitting at a desk and in front of a computer screen, then we are reshaping reality. Conversely, we need to reshape information architecture to better serve us and our changing needs. A huge challenge indeed, but where there is a challenge there is an opportunity. If anything, we do not think this new information architecture is big, or little: we think pervasive information architecture is broad.

RESOURCES

Articles

Boersma, P. (2004). T-Model: Big IA Is Now UX. *[BEEP]*, November 6. http://beep.peterboersma.com/2004/11/t-model-big-ia-is-now-ux.html.

Hinton, A. (2008). Linkosophy. In *Proceedings of the 9th Information Architecture Summit*. Miami, April 10-14. ASIS&T. http://www.slideshare.net/andrewhinton/linkosophy-355763.

Inalhan, G., & Finch, E. (2004). Place Attachment and Sense of Belonging. *Facilities*, 22(5/6), 120–128.

Klyn, D. (2009). Conversation with Richard Saul Wurman. *Wildly Appropriate*. http://wildlyappropriate.com/?p=781.

Leganza, G. (2010). Topic Overview: Information Architecture. *Forrester Research*, January 21. http://www.forrester.com/rb/Research/topic_overview_information_architecture/q/id/55951/t/2.

Resmini, A. (2010). Of Patterns and Structures. *Andrea Resmini blog*, October, 13. http://andrearesmini.com/blog/of-patterns-and-structures.

Weiser, M. (1991). The Computer for the 21st Century. *Scientific American*, Special Issue on Communications, Computers, and Networks, September. http://www.ubiq.com/hypertext/weiser/SciAmDraft3.html.

Books

Greenfield, A. (2006). *Everyware: The Dawning Age of Ubiquitous Computing.* New Riders Publishing.

Lloyd, V. (2007). *Service Design.* TSO.

Morrogh, E. (2003). *Information Architecture, an Emerging 21st Century Profession.* Prentice Hall.

Morville, P. (2005). *Ambient Findability.* O'Reilly.

Soddu, C., & Colabella, E. (1992). *Il progetto ambientale di morfogenesi.* (Environmental Morphogenetic Design). Leonardo.

Wodtke, C. (2002). *Information Architecture: Blueprints for the Web.* New Riders Press.

Wurman, R. S. (1997). *Information Architects.* Graphis Inc.

Wurman, R. S. (2000). *Information Anxiety 2.* Que.

Video

Xerox Parc Labs. (1995). *Ubiquitous Computing.* http://www.ubiq.com/hypertext/weiser/quicktime/UbiCompIntro.qt.

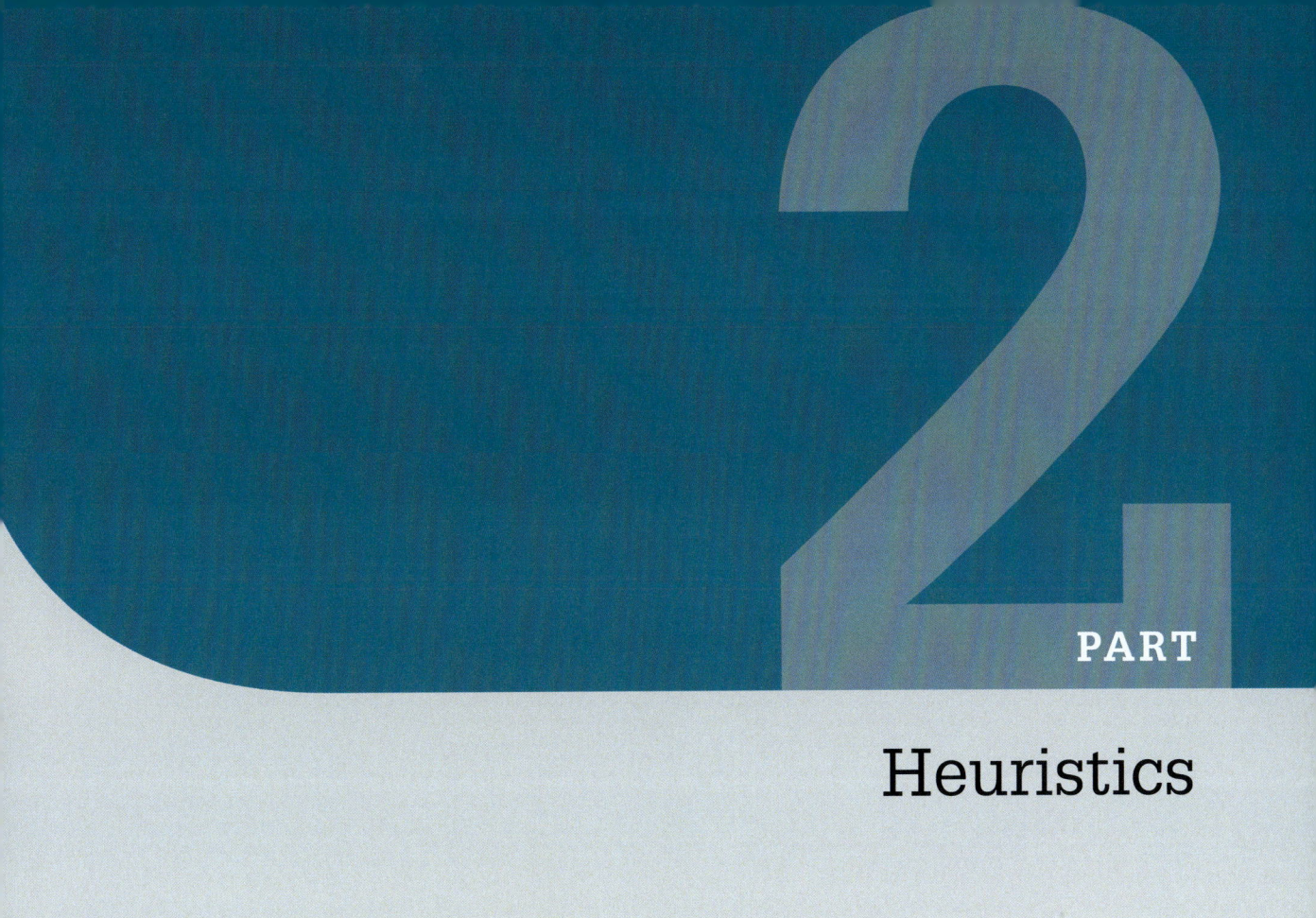

PART

2

Heuristics

CHAPTER 3

Heuristics for a Pervasive Information Architecture

FIGURE 3.1
Photo: Jim G. *Source: Flickr.*

BUILD-A-BEAR

Copenhagen, Denmark, is famous all over the world for its high quality of life, its parks, and last but not least for H. C. Andersen's *Little Mermaid*. Denmark being Denmark and Danes being Danes, after all, it also sports a lively club and bar scene, and if you are into beers, you can tour a couple of very respectable breweries, learn more than you want to know about malt, and enjoy ales aplenty.

But the city has something for everyone, and it can be the unexpected: for example, if you are into plushes, you are in for a real treat at the Build-A-Bear plush store.[1] Now, you find plushes almost everywhere, and you also have specialized shops. Some of these offer limited-availability brands and some go mainstream with the usual armies of cutesy animals or cartoon characters. Build-A-Bear, just

---

[1] Build-A-Bear is an international franchise with headquarters in Saint Louis, Missouri, and you can find their workshops all over the United States, Europe, South Africa, and Thailand.

on the side of the main entrance of the famous Tivoli Gardens on Vesterbrogade, is different. If you walk in, you have to work. Build-A-Bear does not sell you a premade product (a plush), but an experience, a service where children (supposedly, but the shop is quite liberal on this) are the main actors, together with their parents, friends, and shop assistants, and where they happen to build their own custom furry friend. Once inside, children need to do the following:

1. Choose their plush character among a rather large number of different animals and dolls.
2. Plump up the same plush with the help of a dedicated shop assistant and a pumping machine straight out of a movie by Federico Fellini. The machine simply injects feathery stuff into the plush, and children have complete control on how stuffed the plush has to be.
3. Fetch a cloth heart, rub it on their noses, head, and fingers, following a ritual meant to infuse the plush with feelings and life, and then choose what he or she will be best at: reading, roaring, running, roaming, you name it.
4. Insert the cloth heart inside the plush, which now has a life of its own. The shop assistant quickly sews in a few final expert stitches, sealing the heart (and a barcode) inside, checks the plush, and hands it over with a goodbye.
5. Go to a computer that reads the barcode for that specific plush, give it a name, register it, and print its birth certificate.

It's not over though: dresses and accessories are, of course, available to make that plush even more personal; birthday parties can be arranged for both kids and plushes; and rescues can be organized should the fluffy animal or doll (heaven forbid) get lost: the barcode allows one to identify and return[2] lost and founds to their legitimate and distressed owners. "Build-A-Bear is not a shop, it's a workshop!" is their motto. Build-A-Bear does not sell products, it sells experiences, say we.[3]

DANCING WITH USER EXPERIENCE

Un, dos, tres, cuatro: ¡Tierra, Cielo!/Cinco, seis: ¡Paraíso, Infierno!/Siete, ocho, nueve, diez: /Hay que saber mover los pies./En la rayuela o en la vida/vos podes elegir un día./¿Por que costado, de que lado saltarás?

(Gotan Project 2010)[4]

---

[2] In case you are wondering, Build-A-Bear is a privacy-aware enterprise: it is perfectly feasible not to insert the barcode inside the plush as well as possible to walk out of the shop without registering it. This of course means that the animal or doll cannot be tracked at all, for good and for worse. No fears of Big Brother watching us, but no safeguards in case the plush goes AWOL.

[3] If you think plushes are but a small, marginal thing, think of the impact the entertainment industry has on global culture with its gadgets, toys, video games, and movies.

[4] In English: One, two, three, four: Earth, Sky! Five, six: Heaven, Hell! Seven, eight, nine, ten: You have to know how to move your feet. Playing hopscotch or in your life, you can choose one day. Which way, which side will you jump?

FIGURE 3.2
Rayuela, Gotan Project.
Source: YouTube.

Our day-to-day activities are changing. They are becoming cross-channel experiences that require us not only to move from medium to medium, from device to device, but across domains: something that starts digital, such as an e-mail telling us that a product we were waiting for is now on sale, ends up being physical, with us picking it up at the retail store. Or it could be the reverse, with something being shipped or sent to our address, even an electronic address, after a visit to an office.

Gotan Project - The Gotan Project is a Paris-based ensemble that blends Argentinean tango with electronic music, jazz, and a pop attitude. The verse we quote is an excerpt from *Rayuela*, a song from their 2010 CD *Tango 3.0*, whose title and lyrics are taken from Julio Cortázar's novel by the same name. *Rayuela* is the Spanish word for "hopscotch."

In the United States, 53% of consumers reportedly buy products off-line after they research them online, whereas another 43% start their research online, either at their desk computer or through a mobile device, but then find themselves in need to call a customer service number and speak with a human operator to complete the transaction, usually because they cannot find the information they are looking for online (McMullin & Starmer 2010). Similar studies conducted in Europe confirm a strong correlation in consumer patterns among television broadcasting, the mainstream press, and use of the Internet: more than 50% of visitors to online search engines were looking for information related to products or services they saw either in TV commercials or in newspaper advertisements (SEMS 2009). Information coming from one medium is cross-checked or enhanced with related information coming from another medium: this might seem a trivial operation, and to some extent of course it is, as we are simply using what means we have available to improve our chances of getting what we want as we always did. Anyway, this is bound to have larger consequences than the simple flipping through the local variant of the Yellow Pages of old: its impact on design is going to be huge. This constant shifting, this moving back and forth between what is digital and what is physical turns every communication into a cross-channel communication and pushes customers toward a holistic and ubiquitous approach to products and services. This means that we, as designers, and our clients, as producers, need to embrace a correspondent holistic approach to providing those same products and services: multiple separated interactions need to become one seamless flow. As Jess McMullin and Samantha Starmer have pointed out,

> the customer is interacting with (the) brand . . . . they don't care about the channel. I'm the same customer in each interaction; the whole of the experience should be greater than the sum of its parts.
>
> (McMullin & Starmer 2010).

Of course this is easier said than done, since, as it is to be expected, different media have long been developing specific languages, rules, and best practices to communicate their content and engage their audiences. In certain cases, think of newspapers—conventions in typography, layout, language, frequency, and distributive channels have had more than 200 years to get to where they are today. Not only that, but they are the hunting field of specialized professionals, researchers, and companies: it only requires plain common sense to understand that difficulties or inadequacies in envisioning and pushing forth a global approach across channels and domains were just to be expected.

Even so, this shift toward cross-mediality gains momentum all the time, and design we must, as the lack of coordination between communicating or mutually supporting channels is bound to affect the whole process. When multiple interactions are designed as unstructured and unrelated, but are in fact perceived as one single experience by the user, as McMullin and Starmer point out, structural gaps and behavioral inconsistencies are common and unavoidable, and the sheer cognitive load and awkwardness of switching back and forth between noncommunicating and apparently diverse touch points hampers the final user experience.

This is why we believe it is necessary to rethink the design process to be pervasive, ecologic, and holistic: every artifact, product, or service is but a part of what we dubbed, in an article we wrote in 2009, a *ubiquitous ecology*, an emergent information-based system where old and new media and physical and digital environments are designed, delivered, and experienced as a seamless whole. The name simply acknowledges that ubiquitous ecologies share a characteristic of pervasiveness with ubiquitous computing, the systemic nature of media ecologies, and the emergent nature of complex systems.

Two Italian information architects, Davide Potente and Erika Salvini (2009), describe how such an approach has been exploited successfully, to a degree, by Apple, and if and how it could be applied elsewhere. Potente and Salvini argue that the Apple Web site and the numerous Apple stores, in addition to the obvious and necessary interface differences, share a common information organization layer (Table 3.1 and Figure 3.3).

Table 3.1 Comparison between Apple Web Site IA and Apple Retail Store IA

Web site	Stores
Home	Posters on the walls with upcoming products previews
Store	Stands/tables showing products with related details
Mac	Area for Mac computers
iPod+iTunes	Area for iPods, iTunes, and Apple TV
iPhone	Area for iPhones
Downloads	Area for applications
Support	Genius Bar for product support

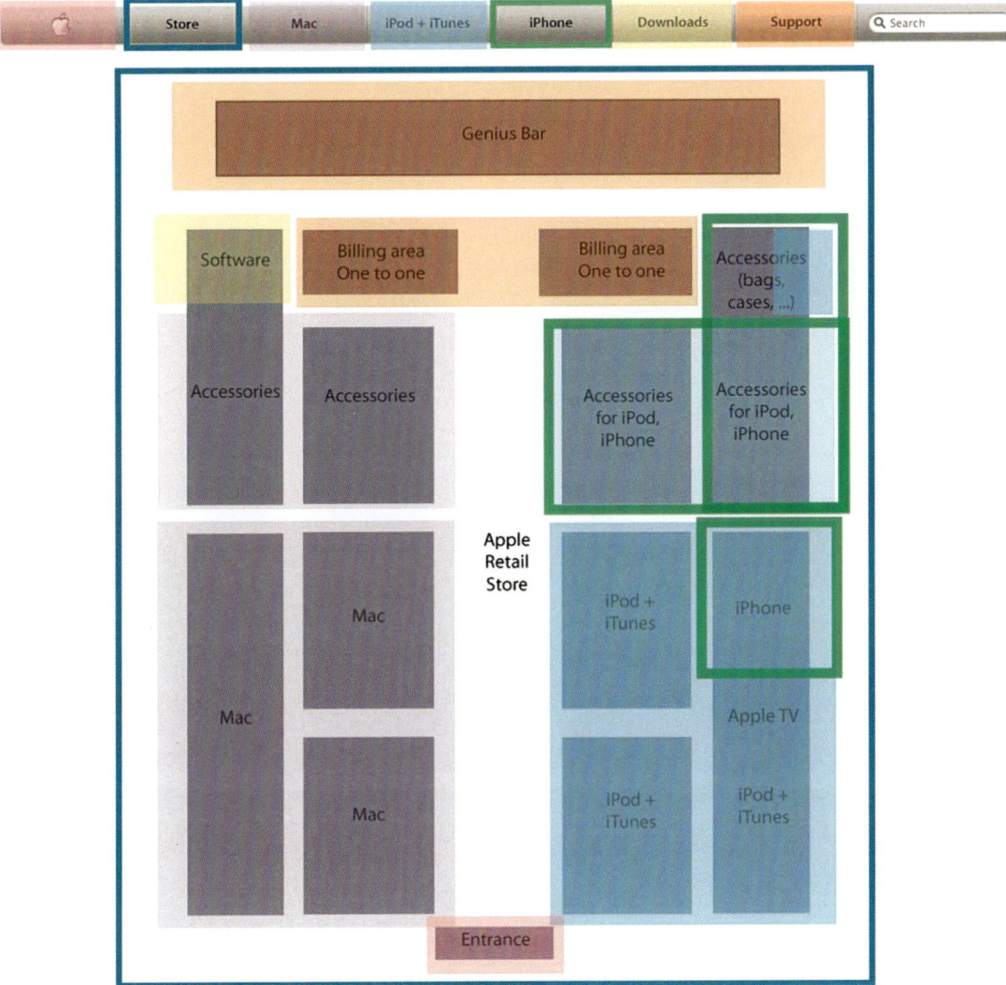

FIGURE 3.3

Map of correspondences between Apple online taxonomy and product placement inside the Apple retail store.

This is true of correlation strategies as well.[5] That is, how information and products are connected works exactly the same both in the Web site and in retail stores. Potente and Salvini (2009) decided to apply these insights to verify user experience at Swedish DIY furniture giant IKEA, as

> customer experience at IKEA is perhaps one of the most representative example of the systemic and pervasive nature of experience. Customers usually start out by exploring and choosing products at home, either on the website or in the paper catalog; then they can move to the store to see them, try them out, and buy them; and finally, at home once again, they assemble the items by themselves following instruction sheets.

With these ideas in mind, they wanted to see if this was reflected in the way IKEA designed and deployed their multichannel strategy as well. They found out right off the start that the paper catalog, the Web site, and the store use radically different information architectures (Figure 3.4).

Menus at IKEA - Being a global giant, IKEA has national Web sites it redirects users to. When Andrea moved to Sweden, it took him a long time to actually be able to use IKEA's Swedish Web site. The reason was that, in the version online at the time, the main local menu listing all super-categories was ordered alphabetically, and of course that order is different in Italian and Swedish. Kitchen becomes *cucina* and *kök*, respectively. As Andrea (thought he) knew the page, he was not actually reading, but homing in to the part of screen where he supposed the item was, and he kept failing for quite some time. This is a good example of how respecting an abstract, absolutely correct principle (follow alphabetic order) in the void might actually hamper the user experience for specific categories of users. IKEA has now changed their approach, and their main lists are the same across all languages: they respect placement, not alphabetic order.

Potente and Salvini set out to develop a single scheme for the entire system, starting from a redesign of the paper product catalog. In the article that documents their design, published in the *Bulletin of the American Society for Information Science and Technology* (ASIS&T), they suggest a number of possible solutions, which range from the strictly information architecture based, such as deploying a single classification structure across the different domains, to the more visually directed idea to codify one single color scheme. What is interesting though is that they suggest a move from multichannel to cross-channel—"to introduce a really transversal information architecture"—with an accent on the fact that some of the principles that work on the Web site should be transferred to physical space. Their idea seems to be that of finding ways to reduce disorientation and increase user engagement through recognition; they have no fear of being bold all over the place. They sketch possible redesigned layouts for the stores themselves, including more ways to access departments in accordance to principles of faceted classification; introduce ways to make customers aware of where they are and where they are going through way-finding techniques and breadcrumbs; organize a totally new taxonomy for the paper catalog; and finally suggest a hub structure for the store (as opposed to the current organic if logical free flow of today) to increase the findability of special spots such as the pick-up areas.

---

[5] We will investigate exactly how this works in Part II, Chapter 8, Correlation.

FIGURE 3.4
Diverging, confusing paths
are like mazes. Photo:
P. Tonon. *Source: Flickr.*

PRECISE AND IMPRECISE

Design: the performing of a very complicated act of faith.

(Jones 1966)

This is by no means a book on design methods, theories, or methodologies. But we feel like a few words are necessary to explain how *we* design pervasive information architectures. It won't take much; you can have some popcorn again meanwhile, and we'll all feel better afterward.

Even though we use the word *process* a lot, we do not see design as an abstract activity; quite the contrary: design is inductive, certainly nonprescriptive, but it's very concrete, directed, and geared toward building both an artifact and a better comprehension of the problem space. Design produces ideas and objects. As such, it needs some base structures in place to be carried out successfully and communicated: this is usually done by adopting methods (you can have more popcorn *now*, in case you already emptied the bowl).

Welsh architect and designer John Chris Jones, in his book *Design Methods*, which we heartily suggest if you want to really grasp this particular subject matter,[6] affirms that a design method is any action one may take while designing, and that regardless of which particular method one decides to follows, there is always

---

[6] The book is an incredible compendium of Jones's own struggle in the field and lists more than 30 full-fledged design methodologies that can be adopted out of the box.

a need for temperance between the opposing forces of intuition and rationalization. This is because design oscillates between creation and communication. Jones exemplifies this dichotomy by structuring the act of designing as the conversation between two sides: *procedure*, "the paper work," and *process*, "the thinking."

According to Jones, procedure is whatever formal superstructure you decide to put in place to assure that certain formal checkpoints can be assessed and anything at all communicated to others. For example, he suggests that at the very least, three logbooks with "data, ideas, and diary of events" will suffice to keep the design flow on track. Process, on the other hand, is design proper and should not be bound by constraint, but wander freely: "the mind must be free to jump about in any sequence, at any time, from any aspect of the problem, or its solution, to another, as intuitively as possible" (Jones 1992). In this view, which is our view as well, the method, albeit important to assure that everyone is on board and that you get paid regularly by your client, is secondary to the "free flow of mind."

One more factor is necessary to consider when assessing what designing pervasive information architecture means. As we move from single artifacts to ubiquitous ecologies, the **gestalt principle**, which is "the whole is more than the sum of its parts," assumes an entirely new, design-oriented meaning[7]: local compromises, or even mishandling and shortcomings, might actually make the global architecture better. Let's try to explain how with an example.

The whole is more than the sum of its parts

In 1967–1971, Dutch architect Herman Hertzberger designed his Diagoon Experimental dwellings in Delft, The Netherlands, in a way to allow its inhabitants to adapt the houses to different, personal configurations within the same underlying structure (Figure 3.5). Hertzberger suggested a number of possible layouts, but the final decisions were left to the individual owners, who could do whatever they liked: they could "literally shape their own environment, while also benefiting from the design help of the architect" (Politano 2006).

This flexibility worked at all scales. Hertzberger used modular concrete blocks, which are manageable and easily handled. As architect and writer Brian Lawson maintains in his really enjoyable book *How Designers Think*, Hertzberger was "far from trying to optimise this object to any one particular function but rather seeing it as sort of compromise." One single concrete block could then be used in the front garden as a "a house number, serve to house a light fitting, act as a stand for milk bottles, offer a place to sit, or even act as a table for an outdoor meal" (Lawson 2005).

---

[7] For more on the gestalt principles in connection with design, see Lidwell and colleagues (2003), especially closure, common fate, figure-ground, good continuation, law of Prägnanz, proximity, similarity, and uniform correctness. For a more general overview, see Gombrich (2000) and Gombrich and colleagues (1973).

FIGURE 3.5
Diagoon housing.
Photo: David Kasparek.
Source: Flickr.

This is an approach that runs contrary to the break-it-down, isolate-the-issues pattern view that scholars such as Christopher Alexander have been pushing. It is also an approach we think suits best the design of complex, user-concerned systems: in designing pervasive architectures it is preferable to sacrifice local details and local precision for a better global experience than it is to do vice versa, as local *imprecision* might become global *precision*. If they are all part of a single process, a somewhat lacking but totally integrated Web site serves its user better than a super-hot, super-trendy Web site that has nothing in common—language, layout, categories, or architecture—with either the paper catalogs or the stores. Breaking things down definitely may help understand, but this understanding does not necessarily imply that we should (re)build complex systems that way.

These points, notes, and concerns are dealt with in detail in Part Three: for now, let's just see how these loose thoughts apply to the design of pervasive information architectures. Just remember that in the design of ubiquitous ecologies, imprecise may not be a bad thing. Not at all.

DESIGNING PROCESSES

Think for a moment of when you go shopping: it is not exactly just buying, is it? It is not the simple act of picking up something from the store (or Web site) and paying for it at the cashier (or on the check-out page). It is much more than that: it starts earlier on, maybe when reading about a certain product in a magazine or seeing it in a commercial on TV, and it might evolve into searching more information about it on the Web or consulting with friends. It might include deciding which retail shop or Web shop is most convenient based on its location, prices, or shipping policies. As Donald Norman says,

> a product is all about the experience. . . . Most companies treat every stage as a different process, done by a different division of the company: R&D, manufacturing, packaging, sales, and then as a necessary afterthought, service. . . . If you think of the product as a service, then the separate parts make no sense—the point of a product is to offer great experiences to its owner, which means that it offers a service. And that experience, that service, comprises the totality of its parts: The whole is indeed made up of all of the parts. The real value of a product consists of far more than the product's components.
>
> (Norman 2009a).

Norman stresses heavily how discovery, purchase, and anticipation (of use) are all a large part of one single user experience. Having just brought what we bought home with us, we still have to open the package, maybe perform some installation, use our newly acquired item for the first time, and check for some assistance. And then maybe it also needs updates, subscriptions, add-ons, changes of plan, or accessories.

At times, anticipation can easily span periods measured in days or weeks, as the item has to travel in from far away, and a whole lot of additional information is pushed our way to comfort, reassure, and engage us while we wait. Track numbers for checking on a Web site, a warning of some delay, which will hopefully have little impact on the final date of arrival, text messages, or e-mail to tell us that port has been reached. So many of these steps are information based and rely on a stream of continuous information: the more this is integrated, the less we feel clueless, stranded, anxious, or plain cheated. This spans the small and trivial—such as using a coherent terminology, fonts, or layout so we don't have to wonder if "dispatch" really means "shipping"–to the big and complex, such as coordinating product interfaces with shop layout as Apple does or coordinating information flows across several different channels.

Design is like Hopscotch - User experience is a large, complex process where designers play a game of hopscotch across the rhizome of multiple channels, contexts, and artifacts.

It's a fact that our experiences with artifacts today are characterized by complexity, instability, and ultimately always **configure a process**, a word that implies an idea of unfolding in space and time of actions, events, or behaviors characterized by a certain continuity. Linguistically speaking, no surprise there: process came to English from the Old French *proces*, originally meaning "journey," and got there from the Latin *processus*, past participle of the verb *procedere*, meaning "to proceed, to move, to go on."

FIGURE 3.6
Interdimensional hopscotch.
Photo: Everfalling. *Source:
Flickr.*

If you remember the quote at the beginning of this chapter, it was the text to a rayuela, that is, a game of hopscotch. It is a children's game that goes back some centuries, one you play by drawing a course you have to jump through alternatively using one foot or two feet (Figure 3.6). That initial quote is not there just because we are fans of the guys in the Gotan Project (we are), but because it allows us to draw some interesting parallelism between the interplay of cross-channel user experience and the complex and multidimensional nature of pervasive information architecture as we intend them in this book. We are sure it is not the worst metaphor you ever listened to, so cut us some slack and read on.

First of all, the original hopscotch is a game that can be played alone or in a group. Then, *Rayuela* is not only a song *on* a game (hopscotch), but it *is* a joke, a play of words and wits in itself, and only a second-level reference to the game. The song is actually an homage and a direct quote of the novel by the same name written by Julio Cortázar: we can hear the writer himself reading some excerpt from his text in the background. Cortázar's novel, in turn, is a hypernovel—a work conceived to allow multiple reading paths and built as an encyclopedia of sorts, loaded with citations from other works and authors, both hidden and explicit: "in its own way, this book consists of many books" (*Table of Instructions*, in Cortázar 1987).

If you are thinking about James Joyce, you are quite on target: *Rayuela*, translated into English as "hopscotch" and published in 1966, is somewhat considered the Hispanic-American literary equivalent of *Ulysses*. It certainly has all the wit, the puns, and the streams of consciousness you associate with Leopold Bloom's wanderings through Dublin, and both texts weave a narrative that draws from complexity. But we also believe it has much in common with the Chinese Encyclopedia Jorge Luis Borges mentioned in his writings and that you will encounter a few chapters down the road, when dealing with the problem of *consistency*. It shares a certain structural resemblance as well with Ludovico Ariosto's epic *Orlando furioso* (and its theatrical rendition by Luca Ronconi in the 1960s–1970s) and with Quentin Tarantino's blockbuster *Pulp Fiction*. These you will meet when discussing *correlation*. Now we are confusing you. That's ok. After all, this is like hopscotch, and you have to follow the flow. As we said before, it's the process, not the procedure. And that's all we have to learn.

A MANIFESTO OF PERVASIVE INFORMATION ARCHITECTURE

To The Inhabitants of SPACE IN GENERAL . . ./This Work is Dedicated . . ./In the Hope that/Even as he was Initiated into the Mysteries/Of THREE DIMENSIONS/Having been previously conversant/With ONLY TWO/So the Citizens of that Celestial Region/ May aspire yet higher and higher/To the Secrets of FOUR FIVE or EVEN SIX Dimensions.

(Abbott 1995)

At this point we expect you to be slightly befuddled. So many different fields, practices, and disciplines are converging into this boundary zone where digital design, networked resources, social interactions, and mobile access blend: why focus on an information-driven approach? Why start thinking about pervasive information architectures?

Well, for one thing, everything is becoming information, and the information we can (and cannot) access increases constantly. In his book *Everything Is Miscellaneous*, American philosopher and technologist David Weinberger (2007) points out how the digital world we are building is not limited in size, scope, and nature the way the physical world is:

in a store, it's easy to tell the labels from the goods they label, and in a library the books and their metadata are kept in separate rooms. But it's not so clear online.

In physical and logical space, what Weinberger calls first order and second order "we've had to think carefully about which metadata we'll capture because the physical world limits the amount of metadata we can make available." Not so in digital space: we can have all the metadata we want. One object, say, a computer, may possess a million pieces of information in metadata. Information is the backbone and is not going away easily: this not only requires attention, it requires design. This is the rationale; then there are some matters of personal preferences.

FIGURE 3.7
Frank Gehry, Der Neue Zollhof, Dusseldorf, Germany. Photo: Problemkind. *Source: Flickr.*

If you allow us one more bad metaphor, there are different ways you can be a brick-and-mortar architect: you could be an architect and build your houses starting from how you want to structure space, such as Frank Gehry does (Figure 3.7) and Frank Lloyd Wright did, or accept the limits of industrial-scale housing and building and work from the inside, such as Giò Ponti or Carlo Scarpa did. Both approaches are legitimate and both qualify as architecture: we just happen to be interested in the Gehry kind of design. We are certainly not saying that this is the *only* way to tackle this issue, but we sure think it is *one* way to do it.

At the same time, designing pervasive information architectures does not translate naively (or purposefully) into a simple enlargement of the playground information architects call their own with a few new hot topics and areas: this is not an exercise in land grabbing. More buzzwords on the business card are not the point: designing artifacts from a structural, informational point of view as the complex open systems they are becoming is it.

We are not in this alone, of course, not at all: we have plenty of company, we just use many different labels and names and come from different places. Donald Norman (2009a), for example, calls this approach *systems thinking:*

> no product is an island. A product is more than the product. It is a cohesive, integrated set of experiences. Think through all of the stages of a product or service—from initial intentions through final reflections, from first usage to help, service, and maintenance. Make them all work together seamlessly. That's systems thinking.

Mike Kuniavsky (2010), one of the founding partners of the UX firm Adaptive Path, calls it simply *ubiquitous computing user experience design:* in his book *Smart Things* he frames devices as *"service avatars"* within a hierarchy of experience scales (covert, mobile, personal, environmental, architectural, urban), embracing and extending the conceptual framework of the service design. User experience, service design, and ubiquitous computing are all coming up to the same crowded intersection downtown where everyone is trying to figure out where (and how) to go next. We call this *next* pervasive information architecture, the design of information within ubiquitous ecologies, and it is definitely interesting to see that, as Peter Morville commented on his Web site,

> while Kuniavsky advises that we view information as one of many design materials (like wood and carbon fiber) from which devices can be made, he also highlights its role as "the core material in creating user experiences."

> (Morville 2010).

Since information architecture relies on principles that are largely independent from any specific medium - after all it is concerned with the structuring of information space as much as architecture is concerned with structuring physical space - it provides a flexible but solid conceptual model for the design of cross-context and cross-channel user experiences which span different media and environments (Figure 3.8). By addressing these structural issues, it is capable of providing all actors with a constant, coherent cognitive framework throughout the whole process. It is important to emphasize that this is not interface design or interaction design. These are both valuable and necessary pieces of the general picture, but they are usually concerned with single touch points, one at a time.

When we say that information architecture needs to structure the process, we move one step up the ladder of abstraction, where information architecture is less of a specific set of tools for, say, Web design and more of a design connector between channels and contexts.

This in turn requires a change in perspective, as it implies that information architecture has to sprout new branches and twigs from its roots and grow taller, richer, and greener. We IAs have to have a little more Wurman in our pockets and move beyond the Polar Bear Book: as information bleeds out to mobile devices and physical spaces, information architecture is not just for the World Wide Web, but helps design all shared informational spaces, places, services, and processes that render the user experience possible in the first place. How can we do this? First things first: we need to acknowledge a few new facts. That's what the manifesto is for. It goes like this.

1. **Information architectures become ecosystems.** When different media and different contexts are intertwined tightly, no artifact can stand as a single, isolated entity. Every artifact becomes an element in a larger ecosystem. All of these artifacts have multiple links or relationships with each other and have to be designed as part of one single seamless user experience process.

Organizations are channel-bound. Customers aren't.
This outlines components and practices necessary to deliver great
customer experiences across more than a single channel.

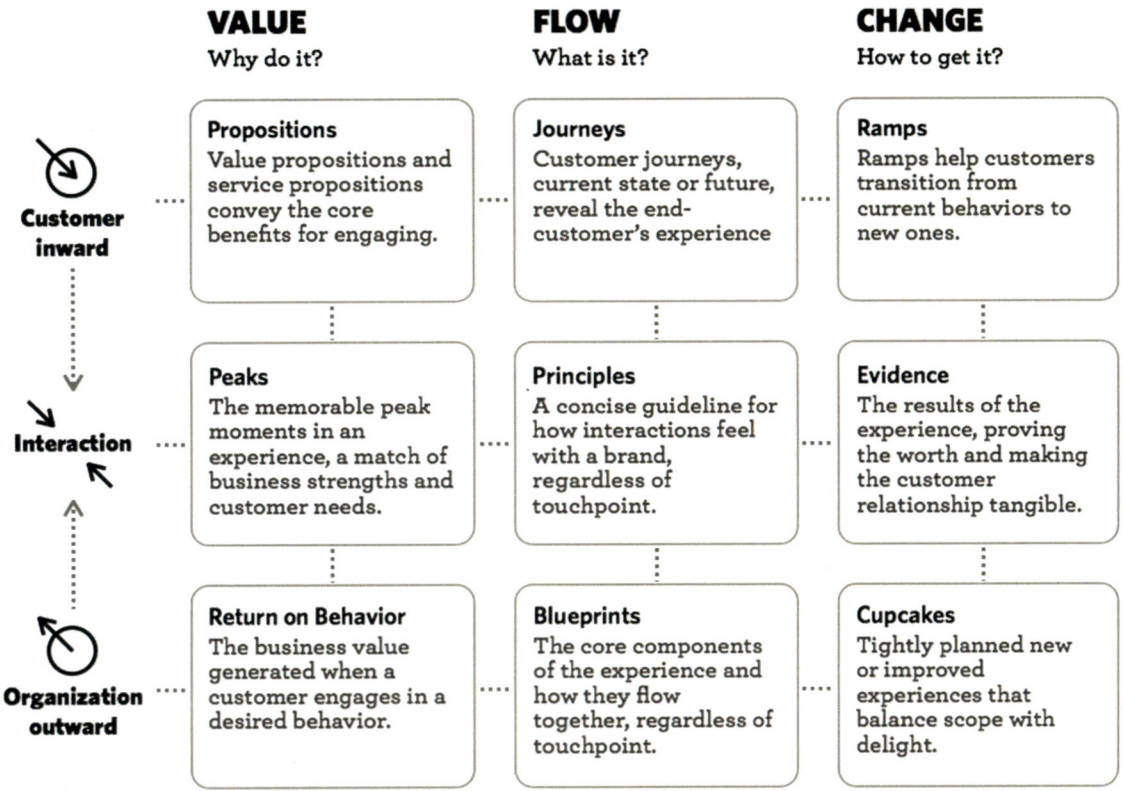

FIGURE 3.8
A take on cross-channel
processes and touch points.
B. Schauer, Adaptive Path.

2. **Users become intermediaries.** Users are now contributing participants in these ecosystems and actively produce new content or remediate existing content by ways of linking, mash-ups, commentary, or critique. The traditional distinction between authors and readers, or producers and consumers, becomes thin to the point of being useless and void of all meaning.

3. **Static becomes dynamic.** On the one hand, these architectures aggregate and remediate content that physically might reside elsewhere and that might have been released for completely different purposes. On the other hand, the active role played by intermediaries makes them perpetually unfinished, perpetually changing, and perpetually open to further refinement and manipulation.

4. **Dynamic becomes hybrid.** These new architectures embrace different domains (physical, digital, and hybrid), different types of entities

(data, physical items, and people), and different media. As much as the boundaries separating producers and consumers grow thin, so do those between different media and genres. All experiences are bridge- or cross-media experiences spanning different environments.

5. **Horizontal prevails over vertical.** In these new architectures, correlation between elements becomes the predominant characteristic at the expenses of traditional top-down hierarchies. In open and ever-changing architectures, hierarchical models are difficult to maintain and support, as intermediaries push toward spontaneity, ephemeral or temporary structures of meaning, and constant change.

6. **Product design becomes experience design.** When every single artifact, be it content, product, or service, is part of a larger ecosystem, focus shifts from how to design single items to how to design experiences across processes. Everyday shopping does not concern itself with the convenience store or supermarket only, but configures a process that may start on traditional media, include the Web, proceed to another shop to finalize a purchase, and finally return to the Web for assistance, updates, customization, and networking with other people or devices.

7. **Experiences become cross-media experiences.** Experiences bridge multiple connected media and environments into ubiquitous ecologies, a single unitarian process where all parts contribute to one global seamless user experience.

HEURISTICS FOR A PERVASIVE INFORMATION ARCHITECTURE

The manifesto outlines what we believe are some relevant trends: to make them into a design method, we had to turn them into actionable goals. At the same time though, we didn't want to overdo it and overrationalize what is at its core an art-and-craft vision: years of practice have just showed us that sometimes making it up as you go is actually what you should be doing – as Eric Reiss (2010) summed up brilliantly in an article for the online design magazine Johnny Holland.

If you are thinking that we or Reiss might be pushing it a bit too far in an effort to score a goal, it might help our discussion to say that this idea is not much of a minority stance among design researchers. Quite the contrary. As Brian Lawson observes in *How Designers Think*, "the comfort of a set of principles may be one thing, but to become dominated by a doctrinaire approach is another." Design, says Lawson, is essentially experimental, and methods, theories, and philosophies are far from being defined precisely, even when their proponents seem to think, or strongly maintain in print, otherwise.

So, if those in the manifesto are the goals, how do you go about them with a hands-on perspective? We retraced our steps and, based on our own experience with the initial stages of design, saw that we could reduce many different, preliminary observations and requests to a smallish set of primary modelers, or heuristics. **Heuristics** are not precise, formalized procedures: they are guidelines, problem-solving suggestions, and directions, not directives.

Heuristics

Heuristics reside in the process part of design, not in the procedure. They are hopscotch, not rocket science.[8] With a few of these, we could identify and impact on process-wide indicators that affect the general design of a pervasive experience, yet retain the "make it up as you go" freedom that is necessary when you are not in familiar waters. After a few iterations, many coffees and muffins, some articles, more than a couple of bored friends, one book, and a dozen public talks, we had a satisfactory lineup consisting of five heuristics and their poignant definitions.

1. **Place-making**—the capability of a pervasive information architecture model to help users reduce disorientation, build a sense of place, and increase legibility and way-finding across digital, physical, and cross-channel environments.
2. **Consistency**—the capability of a pervasive information architecture model to suit the purposes, the contexts, and the people it is designed for (internal consistency) and to maintain the same logic along different media, environments, and times in which it acts (external consistency).
3. **Resilience**—the capability of a pervasive information architecture model to shape and adapt itself to specific users, needs, and seeking strategies.
4. **Reduction**—the capability of a pervasive information architecture model to manage large information sets and minimize the stress and frustration associated with choosing from an ever-growing set of information sources, services, and goods.
5. **Correlation**—the capability of a pervasive information architecture model to suggest relevant connections among pieces of information, services, and goods to help users achieve explicit goals or stimulate latent needs.

We consider place-making, consistency, and resilience to be some sort of ground heuristics, those on which we build. They give the design some anchoring points. Reduction and correlation bring both purposefulness and complexity to the process; they refine, restrict, and expand where and when necessary. They give the design depth.

---

[8] Although, of course, you need some mathematics to be able to play hopscotch and to fire your rockets. The amount required varies a little, though.

At this point, there is a legitimate and pertinent question you may feel like asking: namely, where do these heuristics come from? Not as a group, but individually. Why *place-making* and not, say, *remapping*, *minimization*, or some other cool name?

Unfortunately, as it sure would have been a classy *coup de théâtre*, we have no superior call to support our choices. We cannot claim we received them in our sleep from a spaceship orbiting Venus, nor did we more prosaically come home from a deranged evening in a Chinese restaurant with far too many fortune cookies in our pockets. These heuristics are simply the result of many years of professional and scholarly practice in the field of information architecture. The projects we worked on, large and small, brought us to reflect on our own design practice in the frame of a wider perspective – one capable of moving beyond a single product- or artifact-oriented approach (be it a Web site, a coordinated corporate project, or a physical/digital installation) to embrace a more holistic or ecological approach that wouldn't leave us out in the cold as our projects moved into cross-channel territory. These are the heuristics that allow us to impact on those specific issues in the design of cross-channel ubiquitous ecologies we think are strategic for its success.

We claim no illumination; we claim no splendid isolation: as said before, so many brilliant thinkers and designers are working on these issues. Sometimes we wake up from dreams where we are attending some grand information architecture opening night in our underwear, a rubber duck and flippers, and our set of heuristics.

But the truth is, designing is like playing hopscotch, and these guidelines are essentially the result of our own personal design journey so far, a qualitative, bottom-up process, subsequently integrated by those personal reflections (over coffee and muffins) and long discussions with (bored) friends and colleagues in the quiet of our houses or in the most improbable places in the world we mentioned, with lots of further readings. They work well enough, and we pass them on to you for discussion, not for worship. Nothing in design is worth worshipping; everything is worth a try.

Lawson (2005) writes that "some designers seem to see their whole career as a journey towards the goal of ultimate truth, whereas others seem more relaxed and flexible in their attitudes to the driving forces behind their work": we definitely belong to this second group.

We did some comparative analysis as well, not as a mere, solipsist exercise in style, but as a useful check on the validity of some of our initial assumptions and to examine connections, interactions, or latent influences from preexisting authoritative frameworks as thoroughly as possible. There are many of these, and they range from S. R. Ranganathan's *Prolegomena* and *Five Laws of*

Library Science to Edgar Morin's ideas on complexity, from Celestino Soddu's seminal ideas on *morphogenetic design*[9] to Micheal Graves's **postmodernism**, and, of course, from ubiquitous computing to *everyware*. You'll find echoes of all of these and more as you read along, and we will expand and comment whenever necessary or provide references. But since ubiquitous computing is probably the most obvious connection you have certainly heard of, it deserves a few more lines to clarify in which sense the idea of pervasive information architecture is different.

> **Postmodernism** - A reaction to modernism and to its scientific allure of objectivity, rationality, and progress, postmodernism is a tendency in contemporary culture that spans from architecture to philosophy, which rejects objective truth and the possibility of a single, global narrative. It often emphasizes self-conscious citationism and the reuse of pop patterns, motifs, and memes in acculturated contexts.

UBIQUITOUS COMPUTING AND EVERYWARE

At the heart of ubiquitous computing is the idea that information is processed all around us in all sorts of everyday objects and activities for our use and consumption: it is a system-oriented vision where a constellation of closely related, participating items bridges atoms and bits. And that, in turn, bring us back to the concept of open, self-organizing, complex information systems, what we dubbed ubiquitous ecologies. Ambient intelligence, Adam Greenfield's everyware (Greenfield 2006), and Peter Morville's pioneering ideas on ambient findability (Morville 2005) all prefigure the necessary adoption of a new holistic vision in the design of information spaces as much as a radical change in the way we experience our interactions with information:

> the stakes, this time, are unusually high. A mobile phone is something that can be switched off or left at home. A computer is something that can be shut down, unplugged, walked away from. But the technology we're discussing here—ambient, ubiquitous, capable of insinuating itself into all the apertures everyday life affords it—will form our environment in a way neither of those technologies can. There should be little doubt that its advent will profoundly shape both the world and our experience of it in the years ahead.
>
> (Greenfield 2006, p. 6).

Adam Greenfield's basic assumption is that "information processing is dissolving in behavior" and his observations on *everyware* are built loosely around 81 theses, brief enunciations that highlight characteristics of these new spime

---

[9] See Soddu and Colabella (1992). A glimpse of Soddu's work can also be seen at http://www.argenia.it/.

ERIC REISS—THE MANIFESTO IN PERSPECTIVE

For me, information architecture could just as well be called "thing contextualisation." The "thing" doesn't necessarily need to be information, and the context certainly doesn't need to be on a computer screen. Ultimately, it is the good (or bad) arrangement of "things" within a specific context that creates a specific user experience. And creating the desired user experience must, after all, be the goal of our work with pervasive information architecture. Let me tell you a story to illustrate my point. But first, some background.

Welcome to the Danish Royal Theatre

Originally, I trained in the United States as an actor and stage director. In 1976, I was invited to become the assistant to the famed Danish director Sam Besekow at the Danish Royal Theatre in Copenhagen, Denmark. Sam himself had been an assistant to the legendary Max Reinhardt, head of Deutsches Theater in Berlin back in 1930. Naturally, I jumped at this chance to become part of an extraordinary creative lineage.

One of our first projects together was *Saturday, Sunday, Monday*, a play by the Neapolitan playwright Eduardo de Filippo. Our set designer was Helge Refn, also a theatrical legend. Our 17-member cast was a virtual who's who of Danish theatre (Figure 3.9).

Saturday, Sunday, Monday is a three-act comedy featuring the extended family of demanding matriarch Mama Rosa and her blustering husband Peppino. During the first act, Saturday, we meet the family in Rosa's grand Neapolitan kitchen and are introduced to the basic dramatic conflicts while the traditional Sunday meal is being prepared. The second act, Sunday, takes place around the dining table where more family skeletons in the closet are revealed and everything ends in chaos. The final act, Monday, resolves things amicably, as they do in these kinds of "well-made" plays.

After 5 weeks of hard work, we moved out of the rehearsal hall and onto the main stage where our sets were waiting.

Italy Meets Denmark

The "Old Stage" at the Danish Royal Theatre features the baroque ornamentation that places our disreputable business in the same league as churches—architecturally if not always spiritually. Helge's sets were exquisite—naturalistic, yet with uniquely expressionistic overtones that taunted the imagination. He told me, "Everyone thinks they know what a big Italian kitchen looks like. But I went to Naples and did the research. I'm providing exactly enough reality to point our audience in the right direction, without making them actually question choices I have made that run counter to their own fantasy." Helge was a very wise man.

Despite the magnificent sets, the run-through of the first act was lackluster. Some of this could be excused by the change of venue. However, by the second act, Sam was clearly peeved. "I want real wine in those glasses!" he ordered. Of course, drinking actual alcohol was against some unwritten Royal Theatre rule. But Sam had the clout to get things his way. Real wine appeared at rehearsals the next day—in a surprisingly decent quality.

Sam gave me an important lesson in direction over coffee that afternoon. "The actor is the centre of art on the stage. Give an actor a glass of coloured water, and he will look like an actor *acting* the part of someone drinking wine. But give him *real* wine, and he will look like someone drinking wine—he doesn't have to *act* and can therefore concentrate on his art. Our job is to help actors concentrate on the important things." Basically, this is Sam reducing mental clutter to enhance *Chi*. Feng shui meets Max Reinhardt.

Back to the First Act

The play was taking shape, but things were still very rocky during the key first act. The actors were trying too hard and it showed. Sam was grouchy. Helge was depressed. During dinner that night (at a bad table in the rear of an Italian restaurant), I had an epiphany.

Let me share the first lines of the play with you:

> ROSA: Haven't you finished yet?
>
> VIRGINIA (maid): Nearly. Only two more.
>
> ROSA: Hurry up—I'm waiting.
>
> VIRGINIA: Signora, I think I've done enough already.
>
> ROSA: Are you telling me how to make *ragu*? The more onions there are, the thicker the sauce. I'll tell you how to make *ragu*, it's all in the cooking. Slowly, over a low

ERIC REISS—THE MANIFESTO IN PERSPECTIVE—CONT'D

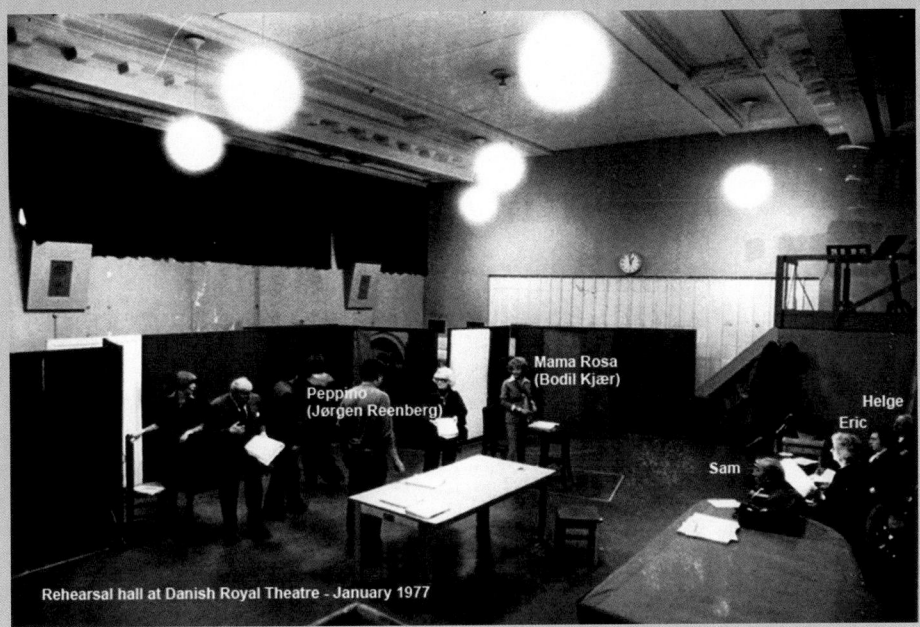

Peppino
(Jørgen Reenberg)

Mama Rosa
(Bodil Kjær)

Helge
Eric

Sam

Rehearsal hall at Danish Royal Theatre - January 1977

FIGURE 3.9
Rehearsal hall at Danish Royal Theatre, January 1977. Photo courtesy of E. Reiss, photographer Aage Sørensen, Nordisk Pressefoto A/S.

flame. Then the onions curl up round the meat in a black crust. When you add the white wine, the crust loosens. That makes a rich golden stock and then you mix it with the tomato sauce and that gives it that lovely dark colour. *Ragu* shouldn't only taste right, it should look right. Don't you tell me how to make *ragu*!

The next day, I got to the theatre early. I rounded up Helge and our stage manager, V. P. Schmidt. I told them my plan. Within the next hour, we'd fitted working hotplates to the prop cast-iron stove. We'd installed a fan at the back of the set (Italian kitchens can get hot, right?). And we'd bought a sack of onions and a sharp knife (as opposed to the blunt props we usually trust to actors).

A Kitchen to Remember

Well, neither Helge nor I told Sam about this change. We just did it. The rehearsal started and, suddenly, the theater was filled with the smell of frying onions. The actors reacted (only the actresses playing Rosa and Virginia had been told). Those of us sitting in the darkened auditorium reacted. You could *hear* the onions, too. We were making *ragu*, not just pretending.

The act finally came together.

On opening night, the curtain went up at 8:07. By 8:20, tummies were rumbling. At the interval, the snack vendors sold out. By the way, the reviews the next day were great. It *was* a great show, thanks to a great script and superb acting. Sam, Helge, and I merely gave our talented cast the means to an end.

Why I Told You This Story

I think this particular theatrical anecdote is interesting because it encompasses the key sensory and cerebral interactions that define the quality of a "user experience." As you read through the rest of this book, I hope you'll pause a moment and reflect

Continued

ERIC REISS—THE MANIFESTO IN PERSPECTIVE—CONT'D

on how the points in the Pervasive Manifesto can be mapped directly to the events I've just described. You'll be surprised at how easy it is and what things you need to be thinking about when you do your next design—on-screen or on-stage.

Born in San Antonio, Texas, in 1954, Eric Reiss has held a wide range of eclectic jobs from ragtime piano player (in a St. Louis house of ill-repute) to senior copywriter (in an ad-house of ill-repute). Eric wrote Practical Information

Architecture, is responsible for Web Dogma '06, and was a cocreator of the world's first "Slam" design competition. He is a former two-term president of the Information Architecture Institute and a professor of usability and design at the IE Business School in Madrid, Spain. Eric is CEO of the FatDUX Group, headquartered in Copenhagen, Denmark, with offices and representatives throughout Europe and North America.

contraptions, as Bruce Sterling would probably call them, and of the social and cultural changes they bring along, one at a time. Some of these theses describe ideas that are pretty close to the five heuristics:

> *Thesis 21.* Everyware recombines practices and technologies in ways that are greater than the sum of its parts.
>
> *Thesis 22.* Everyware is relational.
>
> *Thesis 31.* Everyware is a strategy for the reduction of cognitive overload.
>
> *Thesis 40.* The discourse of seamlessness effaces or elides meaningful distinctions between systems.
>
> *Thesis 41.* Before they are knit together, the systems that comprise everyware may appear to be relatively conventional, with well-understood interfaces and affordances. When interconnected, they will assuredly interact in emergent and unpredictable ways.
>
> *Thesis 47.* The practice of technological development is tending to become more decentralized.
>
> *Thesis 49.* Present IT development practice as applied to everyware will result in unacceptably bad user experience (Greenfield 2006).

These observations largely apply to how we see pervasive information architectures. They are complex systems where the sum is more than its parts and that rely heavily on relationships: *how* is far more important than *what*. *Correlation* is the heuristic indicator that addresses this quality. Similarly, pervasive information architectures are evolving, unfinished, unpredictable systems, or by any means not entirely predictable. This is because, as much as open systems, such architectures are dynamic, undergoing perpetual changes under the actions and influx of people, time, and context. This is what *place-making* and *resilience* try to capture and address.

Ultimately, what differentiates the two is the approach, and the goal: Greenfield's *everyware* is a theoretical framework that tries to explain a trend, a general phenomenon of convergence supported by mobile and ubiquitous computing in general terms. Pervasive information architecture is a heuristic methodology that focuses on the design of the information flows that underlie ubiquitous ecologies. We want to design the damn thing.

How to structure these new complex, compound artifacts via a heuristic process by means of place-making, consistency, resilience, reduction, and correlation is going to be the core of this section of the book. We explain and explore these in depth, one at a time, in the following chapters: we briefly introduce the theme and issues being addressed, mainly in the form of storytelling; present and discuss the heuristic itself; introduce a series of case studies spanning the physical and the digital; recap briefly, in the form of bullet-point lessons that can be applied while designing; and finally finish you for good with a bibliography of relevant articles, books, movies, videos, or games. Are you ready? Then hold tight, the ride is about to begin. We cannot promise that there will be no bumps and a few scary moments, but we can certainly promise it's going to be fun. We might even get to use some of those map-making, sword-fighting, code-writing skills we learned on the Internet. Here we go.

> **Rosen** - Rosen (1999) offers an interesting, relational view on complexity. A system is "simple if all its models are simulable. A system that is not simple, and that accordingly must have a nonsimulable model, is complex." In other words, Rosen ties his concept of complexity to modeling, where this is the act of establishing congruence between the elements and the structures of two systems, the one being observed and its model. A simple system can be simulated and predicted; a complex system cannot: "When a single dynamical description is capable of successfully modeling a system, then the behaviors of that system will, by definition, always be correctly predicted. Hence, such a system will not have any 'complexity' in the sense above, in that there will exist no unexpected or unanticipated behavior."

RESOURCES

Articles

McMullin, J., & Starmer, S. (2010). Leaving Flatland: Designing Services and Systems across Channels. In *Proceedings of 11th Information Architecture Summit.* Phoenix, April 9-11. http://2010.iasummit.org/talks/9702; http://www.slideshare.net/jessmcmullin/leaving-flatland-crosschannel-customer-experience-design.

Morville, P. (2010). Ubiquitous Service Design. *Semantic Studios*, April 19, http://semanticstudios.com/publications/semantics/000633.php.

Norman, D. (2009). Systems Thinking: A Product Is More Than the Product. *Interactions, 16*(5), September/October. http://interactions.acm.org/content/?p=1286.

Reiss, E. (2010). In Defense of "Making It Up as You Go Along." *Johnny Holland Magazine* July 28. http://johnnyholland.org/2010/07/28/in-defense-of-making-it-up-as-you-go-along/.

Resmini, A., & Rosati, L. (2009). Information Architecture for Ubiquitous Ecologies. In *MEDES '09 The International Conference on Management of Emergent Digital EcoSystems,* Lyon, October 27-30. http://doi.acm.org/10.1145/1643823.1643859. (Also available at http://andrearesmini.com/blog/ia-for-ubiquitous-ecologies.).

Books

Abbott, E. A. (1995). *Flatland: A Romance of Many Dimensions.* London. (Available at http://www
.gutenberg.org/etext/201 or http://www.archive.org/details/flatlandromanceo00abbouoft).

Cortazar, J. (1987). *Hopscotch.* Pantheon.

Gombrich, E. H. (2000). *Art and Illusion.* Bollingen.

Gombrich, E. H., Hochberg, J., and Black, M. (1973). *Art, Perception, and Reality.* Johns Hopkins
University Press.

Jones, J. C. (1992). *Design Methods.* Wiley.

Kuniavsky, M. (2010). *Smart Things.* Morgan Kaufmann.

Lawson, B. (2005). *How Designers Think.* Architectural Press.

Lidwell, W., Holden, K., & Butler, J. (2003). *Universal Principles of Design.* Rockport Publishers.

McCullough, M. (2004). *Digital Ground.* MIT Press.

Rosen, R. (1999). *Essays on Life Itself.* Columbia University Press.

Weinberger, D. (2007). *Everything Is Miscellaneous.* Times Books.

Music

Gotan Project. (2010). Rayuela. In Gotan Project, *Tango 3.0.* Ya Basta! Records.

CHAPTER 4

Place-making

FIGURE 4.1
Cambridge, UK.

ANDREA TRAVELS WEST

Perhaps the country only existed in its maps, in which case the traveler created the territory as he walked through it. If he should stand still, so would the landscape. I kept moving.

(Greenaway 1978).

A couple of years ago I had a chance to do some research-related traveling during the summer and I happened to stay in Cambridge, United Kingdom, home of the world-famous Queen's and King's colleges, for a whole weekend. All duties disposed and taken care of, I decided to go into full tourist mode: I took a lot of pictures, visited all the right places, and ate in a half a dozen bad restaurants. Cambridge is a beautiful city, albeit it was so stuffed up with Italians at the time that it looked and sounded like the Riviera.

I walked a lot, and being the resourceful kind of guy I had a foldable paper map with me all of the time. Cambridge is a medieval city, has plenty of monumental buildings at its center, and although the river Cam certainly makes it even more interesting and picturesque with bridges, dams, and all, it sure did not provide the city with the perfect site for laying out a regular street grid: Cambridge is your classic web of turning, winding streets. While cruising St. John's Street I walked into a map of the city (Figure 4.2):

FIGURE 4.2
Street map, Cambridge, United Kingdom.

I looked at it, and I got completely lost. I didn't recognize the city it depicted. I knew it had to be Cambridge: of course it had to, who would place a map of Exeter there, but I couldn't make any sense of it. I wasn't running low on sugars, and I'm pretty good at maps, but I just couldn't read it.

The map per se was your pretty normal, standard "You are here" street map. It was in a visible, accessible place and large enough to be readable even from a few steps away. Typefaces, colors, wording, icons, and everything about it were neither particularly visionary nor plain wrong. It listed major monumental buildings and places, facilities, parkings, and even threw in a few directional arrows for top-of-the-list locations from there.

After being puzzled for the good part of 5 minutes, I took a very thorough look and I saw something I had not noticed at first: in the lower right-end corner there was a neat arrow pointing north. East. Sorry, north, it said, but in the normal way we look at maps, with the north up, it was actually pointing east. I realized the map was simply rotated 90 degrees right (Figure 4.3).

FIGURE 4.3
North is east.

Once I subdued a sudden urgency to kick the panel hard and loud enough to risk an arrest, I was able to turn the map 90 degrees left in my mind and everything fell into place: "Now, here you are. Well, of course. Queen's College, indeed, and there's the Segdwick Museum. Jolly good, jolly good."[1]

---

[1] Note that Andrea does not actually talk like that at all, but this being Cambridge we figured out you could cut us some slack on local color.

BEING THERE

But why was it so? There was nothing less than utilitarian about the map: it was a standard aerial view, very mappish if you wish, and had no particularly uncommon features if you exclude that east–north inversion oddity. The reason became immediately apparent after studying the map those 5 more minutes: it graced with its presence a totally nondescriptive site where no outstanding landmarks were visible,[2] but nonetheless it applied what is generally called *local structure matching*. In other words, even though nothing in the neighborhood was suggesting the possibility of a **visual alignment** between elements in the map and elements in the landscape, the map was designed to show the view as it was from that very point of St. John's Street. Unfortunately, this did not really help at all: since nothing except for that small, unnatural east-pointing north pointer told me I needed to rotate, I read the map as we normally do with all maps, including the foldable one I was carrying around, figuring north was up. This immediately threw me off, and I got lost.

Aligning the map with the world

It was an eerie, unnerving sensation, and for a couple of minutes it felt like I had stepped out of the map of the known world. I knew I was walking around Cambridge, safely enjoying my tourist stroll with all the time in the world, so nothing particularly terrible could happen. But it could definitely have been worse. And not just because it could have been raining. Think of our friend Mr. Jones from our little story in Chapter 1 and think about going through an ever-changing series of maps, navigational aids, and mental models. This constant shifting and adjusting is the required price that we have to pay to be able to carry some complex task to conclusion. It really does not work. There is a basic need for continuity and the creation of a recognizable "being there." This is what **place-making** is about: being there, laying the foundations of a ubiquitous ecology.

Place-making - The capability of a pervasive information architecture model to help users reduce disorientation, build a sense of place, and increase legibility and way-finding across digital, physical, and cross-channel environments.

SPACE, PLACE, AND TIME

We say navigate, but really mean understand.

A few years ago, in a discussion on the news site Slashdot speculating on Keanu Reeves's possibilities to be cast as the new Superman, someone asked if that was not a bad choice, as Keanu *"is half Hawaiian"* and Superman *"is white."* Regardless of Reeves's descent, which is related only fractionally to Hawaii, reader Nightpaw pointed out correctly that "Superman is Kryptonian. He's a different species. He's not a member of any human *race*. We're lucky he's even

---

[2] Actually, it was even a little worse than that, as can be seen in Figure 4.2, as the map was placed right before a gate, which effectively barred the way, both visually and physically.

bilaterally symmetrical."[3] It might sound like a joke in reply to a slightly dumb or even racist comment, but it's the truth. And has huge implications as far as Superman and his empathy and capability to understand us and our world go.

We are physical beings: the very idea of space, of embodiment, shapes **the way we perceive reality**. It's by no chance that we say the future is ahead of us, where our eyes look all the time and where our steps take us, and the future is behind our back, where we cannot see anymore. Our language reflects a long list of spatial metaphors, both positive and negative, that include figures of speech such as to sink into a depression (and the deeper it is, the harder it is to come out of it), to be at the top, to think backward, or to fall from grace. You probably can think of many more. This is a common cognitive mechanism, well studied and well known: we grasp the abstract by means of the concrete.[4] So it shouldn't come as a surprise at all that we brought this language to the Web: we go to a Web site, we go up, down a page or a hierarchy, we exit, we open a link, we close a pop-up. That is, when we needed some way to understand and share the whereabouts of hypertextual space, we resorted to our embodied routines. These have roots so deep, we move around to get objects all the time, that we never seriously questioned the fact that it could be argued that **we are not going anywhere**, it's the Web sites that come to us (Dourish 1999).

Embodiment shapes our perception

On the Web we are not really going places

Saying that we have a perception of reality and a thinking mind shaped by our physical form is far from being deterministic, as there is a lot of individual, cultural, and social variance, but that provides us with a starting point we cannot dismiss, as much as we cannot dismiss the fact that Superman is an alien. If forks or fork substitutes are and have been roughly the same size all around the world, that tells us something about the structure of our hand, the length of the arm, and the size of the mouth. Same goes for your computer keyboard: it wouldn't be too useful to have keys the size of chairs you have to jump on to press (fun, yes, for a while; useful, no) or, as everyone using a mobile phone painfully knows, keys the size of peas you can only hit precisely if you train hamsters to type for you.

What is important here is to understand that most of our experience of space has nothing to do with the idea of **geometry** and of Euclidean measures we learn in school but is strictly related to embodiment. It might seem intuitive, but it's not. Or maybe it's not for a lot of people who should know better (and that includes us as well).

Space is not geometry

French cultural anthropologist and paleontologist Claude Leroi-Gourhan argues in his book *Gesture and Speech* that tools (gestures[5]) and language (speech) are not the most significant human inventions, but their by-product,

---

[3] "Keanu Reeves as Superman." Slashdot, September 14, 2002. http://news.slashdot.org/article .pl?sid=02/09/14/232239.

[4] For more on this, see Chapter 5, Consistency, when we introduce metaphors and metonymies.

[5] It is remarkable that gestures often express emotions as well, considering how "these patterns are at times conscious and coded in language but we are often not aware of the basis of our actions" (Chazan, 2004).

Domestication of space
and time

the **domestication of space and time**, is. Leroi-Gourhan considers gesture and speech equivalent in the two domains, as a gesture is as much a movement of the body as speech is a movement of breath through the larynx, but through the pace and rhythm of walking and speaking, chanting and painting, humans are capable of constructing the very idea of an inhabitable space that would have been otherwise unattainable.

In 1963, German philosopher Otto Friedrich Bollnow published a book entitled *Mensch und Raum*,"[6] *Man and Space*, which never got translated into English.[7] While this circumstance might certainly explain why Bollnow is largely unknown to English-speaking audiences and hence little quoted or directly cited, we nonetheless owe Bollnow a tremendously important new vision of space as an **anthropological** rather than a physical or mathematical concept, with humans at its center. His main three points are that

Anthropological space

1. Space is heterogeneous
2. Space is hodological
3. Space has evolved

That space is heterogeneous simply means that space is no homogeneous expanse. Bollnow introduces a relativistic or subjective point of view, connected to individual experience, which negates any fixed points: all references happen within a subjective system, be it a familiar one, such as when departing and returning to a well-known spot such as home, or an unfamiliar one, such as when checking in at a hotel in a city we have never been before. Central to this view is the idea of an ever changing center of space:

> we move out of our apartment to a new one, our whole world is newly reorganized from the new one. As a consequence, we get to know space environmentally, in a continuous tension between inhabited, well-delimited areas and the surrounding chaos.

Hodological space

Then, space is also **hodological**, from the Greek words *hodos*, path or way, and *logos*, discourse. Bollnow's take is that human space is totally different from mathematical space: it is a space of paths, and experience, and it corresponds exactly to what we perceive if we move between two different locations:

> in contrast to the mathematical concept of space as presented on maps, plans, etc. "hodological space" is based on the factual topological,

---

[6] Apparently the word *raum* comes from the verbal form *räumen*, which means "to clear a part of the wilderness with the intention of settling down, to establish a dwelling."
[7] Christine Shuttleworth is working on an English edition of the book to be published by Hyphen Press under the title *Human Space* in late 2010. At the time of this writing the book is still unavailable (http://www.hyphenpress.co.uk/books/978-0-907259-35-0).

physical, social, and psychological conditions a person is faced with on the way from point A to point B, whether in an open landscape or within urban or architectural conditions. Bollnow gives many interesting observations on the cultural implications of hodological distances as compared and contrasted with geometrical distances (language and culture in mountain valleys; traditional traffic conditions in mountainous regions; the structure of war landscape with its absolute focus on the front).

(Egenter 1992).

Think of an apartment house: two points might be just inches away, but if they are separated by a wall, for example, they are in different units and their hodological relation is going to be radically different from what the architect might envision geometrically, so to say, on the blueprints.

Finally, space has not always been there, it has **evolved**. Bollnow maintains that a universal idea of space is indeed a very late event in history and is largely connected to the age of discovery and cartography of the 15th and 16th centuries. Even so, what we call space is actually a sequence of historically defined ideas of space. We know about the Romans and their quadripartite cities, the gruma pole, and the art of divining a proper place for a new settlement in the flight of birds. We know about medieval maps with Jerusalem at the center of the known world and Noah's Ark landing place close enough you could take the kids on a day camp to see the animals (Figure 4.4).

We know all of this, but we are not worried anymore to fall off the disc of the Earth if we sail to the horizon[8] nor do we resort to the patterns in the flight of a flock of gulls to decide if we are going to buy that house on the outskirts or the one downtown. And wait, we know you might say that trying to understand the real estate market amounts exactly to that, divining, but we are not going into that. We know too many economists.

Space has evolved

> **Gromatici** - The agrimensores or gromatici, from the groma or gruma pole they used, were Roman surveyors who translated the auspices of the augurs into the actual layout for roads, camps, and cities; whether they had a religious role or concern themselves is a matter of some debate. Rome and its colonies were all laid out according to these complex rituals and measurements, which aligned human space with the boundaries of the templum, the consecrated space that was chosen for the new settlement.

Bollnow's ideas were carried over to architecture by Christian Norberg-Schulz, a Norwegian architect and academic. Norberg-Schulz argues that architectural space is a compound of many different layers that include emotions, Bollnow's hodological space, and topological concepts. He calls this **existential space**: a space of relationships, which is personal, immediate, egocentric, and made up

Existential space

---

[8] Unless, apparently, you belong to the Flat Earth Society. Their Web site sports an abundance of literature that explains how it would technically be impossible to stand on the Earth if it were a sphere, as we certainly would be falling off all of the time (http://theflatearthsociety.org/).

FIGURE 4.4

The Hereford Mappa Mundi, circa 1300, attributed to Richard of Haldingham. Jerusalem is at the center, and the ark can be seen slightly left and top. *Source: Wikipedia.*

of more stable archetypes, vicinity, enclosure, separation, continuity, and time, as the way we see our surroundings change constantly.[9] *Dwelling* implies much more than simple shelter, and in his later book *Genius Loci*, the *Spirit of the Place*, Norberg-Schulz (1979) will definitively identify this space where our life occurs as *place*.

Space and place are different

Space and place are two very different, if often confused, concepts: space is the base experience of our embodiment, and it is objective, impersonal, undifferentiated; place, however, involves a particular kind of presence that

[9] For the sake of brevity, you could say that existential space spans the geometrical, the topological, and the hodological.

includes, in addition to physical space, memories, experiences, and behavioral patterns associated with the locale. It is personal, subjective, and communitarian. Place is what we are bringing into cyberspace.

NAVIGATING CYBERSPACE

A few years ago, it seemed like we agreed that cyberspace was all about virtual reality, either the Metaverse kind from Neal Stephenson's *Snow Crash*, with some good sword fighting thrown in, or the *Matrix* movies variant, with awesome zero-gravity kung fu. Today, video games are certainly getting there, and although there are lessons to be learned in the way they approach user experience, it doesn't seem that we are actually too keen on doing anything more in 3D than having some fun: Second Life's slow descent into irrelevance is a testament to that. The thing is, cyberspace is already here. **Information bleeds out of the Internet and into the physical world** through mobile phones, pads, public real-time displays, house appliances, and any sort of connected devices you might think of. In a way, it's still the Metaverse, only it's the other way around.

It's like cyberspace, only the other way around

Paul Dourish has documented how our perception of the way we browse the Web, regardless of how knowledgeable we are, has a lot to do with going through sequences of paths and nodes in a way that cannot but remind the way American urban planner and MIT professor **Kevin Lynch (1960)** described how we experience urban environment in his seminal book *The Image of the City*. After a 5-year field study in Los Angeles, Boston, and Jersey City, Lynch found evidence that we move through cities by forming mental maps of the surroundings, mixing five different base elements: paths, edges, nodes, landmarks, and districts. Think of going to work in the morning: if you walk, you will more or less think of going straight (path) until you reach a certain corner or building (landmark); then you will turn right, walk a little more (path), get to the usual café (node), and move on to your office (node). Everything in between is pretty much a blur, to the point that we might miss or erase out most of the landscape simply because it is irrelevant. Things are not different if you drive or use public transportation.

Kevin Lynch - Kevin Lynch's *The Image of the City*, written in 1960, has been a momentous book in the history of urban planning. Lynch wanted to understand how people navigate urban spaces so he conducted extensive experiments with local residents of Boston, Los Angeles, and Jersey City. He had them draw maps of their surroundings from memory and noted that these were basically mental maps, maps illustrating their own personal view of their neighborhoods, built using five basic elements: paths, such as streets, or bus lines; edges, for example, walls, fences, or shorelines; districts, places with a well-defined local identity; nodes, intersection or meeting places, such as the main squares of Italian cities; and landmarks, visible structures that allow long-distance orientation.

Way-finding - Way-finding tries to understand how people orientate themselves dynamically while moving from place to place: how we are able to walk around and make sense of the surrounding environment, remember paths and places, and generally know where we can fetch that tasty sandwich or how to avoid an unsavory neighborhood. It has its roots in urban studies, cognitive psychology the environmental sciences, and psychology. First applied by Kevin Lynch in the 1960s to understand the way we experience urban landscapes, the concept was then expanded and refined by Romedi Passini in his 1984 book *Wayfinding in Architecture*. Way-finding is an important piece of the theories that try to unravel the complex relationship we entertain with digital interfaces, navigation in virtual environments, and the Web.

Lynch called this dynamic process we use to orient ourselves in physical space and navigate between places **way-finding**: the concept was later expanded to include signage and all those elements that help make the grammar of any given space understandable by Romedi Passini, in his 1984 book *Wayfinding in Architecture*. Because of obvious concerns with information seeking, navigation, and user orientation, way-finding has been adopted by information architecture since the very beginning. Plenty of books and articles have been written on way-finding for the Web, and some of them are listed in the references: we just want to stress how the very idea that we orient ourselves by building a map of paths, landmarks, and nodes in our head is basically just begging to be taken from the concreteness of physical space to the abstractness of digital cyberspace.

CENNYDD BOWLES—WAY-FINDING

Some species are particularly blessed with way-finding ability. Salmon can navigate by the scent of minerals around them. Ants possess natural odometers and skylight compasses. Without such generous physical advantages, humans have relied on vision, spatial reasoning and mental models, and the brain's ability to combine explicit and implicit cues to reach our destination. The mental maps so central to our way-finding success are built up through experience and intuition and through three specific modes of knowledge.

Survey knowledge describes our topological understanding of the environment around us. Through survey knowledge we conceptualize the space as a whole. Generally the survey models we create are hierarchical, with large, general places (cities, say) encoded with smaller subnetworks (neighborhoods or streets). *Procedural knowledge* represents a sequence of actions required to follow a route from A to B. We often rely on procedural knowledge when we plan a route in advance through unfamiliar territory. This saves us from having to build up our survey knowledge, but procedural knowledge is, unfortunately, fragile. If, for example, we ask a stranger for directions but forget their first instruction we remain utterly lost. Procedural knowledge can give the answers to the exam, but that's not the same as knowing the subject thoroughly. *Landmark knowledge* describes our understanding of spatial reference points. Landmarks are typically tall trees, buildings, or mountains so that we can see them from various angles and hence interpolate our position.

Designers can improve user way-finding by bolstering these three types of way-finding knowledge, through artifacts such as maps, signposts, directions, and reference points. Clearly these techniques also transfer excellently to the digital world.

CENNYDD BOWLES—WAY-FINDING—CONT'D

To build survey knowledge of a digital space, designers can expose the information architecture of a system to the user. Look for ways to allow the user to form a clear understanding of the organizational structure of the system and how items within it are likely to be classified. Examples include a site map or menu that clarifies the divisions of a Web site, a map of a virtual world, or page boundaries on a smartphone interface.

Procedural knowledge is often built up by users themselves, particularly if they are disinclined to learn details of survey knowledge. Return visitors learn paths to their desired content but, just like the real-world example given earlier, this knowledge is vulnerable to architectural change or error. Redesigns in particular disrupt users' procedural knowledge,

causing them frustration. If users cannot find their feet by investigating the system's IA, the designer may have to signpost explicitly their desired material.

Finally, global elements offer the equivalent of trees and towers, helping build users' landmark knowledge. A stable reference point allows users to reorient if they become lost and therefore gives users courage to explore, knowing that they can always return to a safe anchor point. Consider, for example, the de facto standard of logo-as-home page-link landmark on Web sites or the home key on a smartphone that offers a single, globally accessible virtual landmark.

Cennydd Bowles is a user experience designer, author, and community evangelist based in Brighton, United Kingdom.

FROM SPACE TO SIGN

What Andrea experienced in Cambridge when first seeing the rotated map is well known in way-finding literature and is commonly described as *being turned around*. When relying on maps and not on direct observation, this being turned around might get even worse, as spatial knowledge derived from maps is normally orientation specific and it seems just natural that without physical evidence it might be even easier to get lost.

In physical space, this can be partially avoided by some *structure matching*, the pairing of known points in the environment with points on the map. It is what we do when we turn our map around to align it with what we currently see from your point of view. Unfortunately, for many this only really works when they are presented with some kind of false perspective map, such as an isometric projection; when we align the map to some emergent visible feature, easily recognizable on the spot; or when **some tool** does its magic for you. This was not the case in Cambridge, if you remember.

> **Way-finding tools** - They are usually categorized in five groups: (1) tools that display the user's current position, such as LORAN, a radio-navigation system; (2) tools that display the user's orientation, such as compasses; (3) tools that log the user's movement, for example, the traditional captain's log aboard a ship; (4) tools that show the user's surrounding environment, such as maps; and (5) guided navigation systems, for example, GPS and signage. That map in Cambridge was a so-called YAH map, category 4, tools that demonstrate the surrounding environment, with some category 5 additions thrown in.

However, this becomes nigh impossible if we try to apply navigational logic and way-finding tricks for moving around in a city to the Web. There is no help coming from our physical embodiment, which is the first reason for having proper place-making: we need to be there somehow. Carefully designed way-finding signals can help users build a mental map quickly, effectively, and with a minimal cognitive load, but how do you orientate yourself dynamically when you have no body to move around?

German-born American psychologists Kurt Lewin, one of the fathers of social psychology, introduced the concept of topological psychology in his works in the 1930s. Lewin wrote that distances in psychological space, the space we perceive with our senses, cannot be measured as they can be in **geometrical space**. Psychological distances may be (or feel, if you prefer) either shorter or longer than their physical counterparts. It is something we all have gone through at least once: if traveling is easy, comfortable, or pleasurable, it seems to go in a blink. When we are on a boring or uninteresting and unfamiliar route, it seems to last forever. The concept is not new to us: Bollnow and Norberg-Schulz propose very similar considerations. However, Lewin adds a very important final point: the distance between one hodological area or region and another is not the shortest path but *"the path of least effort*[10] given the attractive and repulsive valences of the regions making up the space,"* which means that something that is really more distant can *feel* closest, can *be* closest if it calls for less activity, less engagement, and ultimately less choice.

Psychological space is different from geometrical space

The hyperlink as the central element of information space

If information spaces are perceived as nodes and paths, the **hyperlink** is the element that allows for those nodes and paths to exist. Andrew Hinton (2009) wrote in his article *The Machineries of Context* that

> the Web (is) becoming the place of record for conversations, stories and even our identities. And that's because it's such a perfect medium for people to associate, connect, and discover. . . . The hyperlink made this possible.

Hyperlink connections are certainly semantic in nature, as they are built on logical connections and not on spatial proximity, but, as Dourish and Chalmers (1994) observed,

> in these systems, we observe not purely spatial navigation, but semantic navigation which is performed in spatial terms. What is gained here is a naturalness of use based on the everyday familiarity of the physical environment. . . . In spatial navigation, a user will move from one item to another because of a spatial relationship—above, below, outside. In semantic navigation, this movement is performed because of a semantic relationship—bigger, alike, faster—even when that relationship is expressed through a spatial mapping.[11]

---

[10] The principle of least effort is discussed in depth in Chapter 6, Resilience.

[11] Dourish and Chalmers identify a third navigational model that they call *social navigation*: "in social navigation, movement from one item to another is provoked as an artefact of the activity of another or a group of others. So, moving 'towards' a cluster of other people, or selecting objects because others have been examining them would both be examples of social navigation."

This vision is not even spatial if we consider how Dourish tends to equal spatial and geometrical: it's hodological or existential. In another paper written with Steve Harrison, Paul Dourish (1996) goes on to clarify that

> appropriate behavioral framing is not rooted in the properties of space at all. Instead, it is rooted in sets of mutually-held, and mutually available, cultural understandings about behavior and action. In contrast to "space," we call this a sense of "place."

Establishing a sense of place is what we call *place-making*. It is as necessary as proper navigation and way-finding to make any design habitable by its users— even more so in information space, where we lack the comfort of our favorite armchair. After all, this is not the Matrix (Figure 4.5).

FIGURE 4.5
A sense of place where there is no space. The 1950s family fireplace in the *Matrix*, Wachosky brothers.

PLACE-MAKING IN PERVASIVE INFORMATION ARCHITECTURE

Place-making is the capability of pervasive information architectures to help users reduce disorientation and increase legibility and way-finding in digital, physical, and cross-channel environments. In ubiquitous ecologies, an important consideration is to structure all elements of a given user experience process as parts of a continuously flowing place. In a process that bridges a number of different channels and environments, successful place-making is a crucial factor in shaping the overall user experience. Users have to feel at home, to be in context, and this is reinforced if all parts of the process are structured in a way that they belong to one single, common, existential space.

The Institute for the Future (2009) wrote in their report *Blended Reality: Superstructing Reality, Superstructing Selves* that "cyberspace is not a destination; rather, it is a layer tightly integrated into the world around us."

Networked hyperlinks are the raw materials through which we weave this compound reality, and although cyberspace, semantic space, digital space, or the Web do not map precisely or exhaustively to a single physical representation of space, they are existential space, places we spend part of our emotional life in. More than that, they need to be places if they want to provide satisfying experiences: when they are not, we get lost, we do not understand, we feel frustrated, and the experience ultimately fails.

This is what we see happen in successful and unsuccessful social networks, for example: as our interactions become mediated by technology and devoid of physical presence, the value of place gets reinforced. Facebook is a good example: it has been successfully positioning (and promoting) itself as *the* meeting place of choice. Its success lies mostly in the fact that everyone is there; our friends are there. It is simply the hip bar in town, on a much grander scale and without ex-pro footballers turned bouncers at the doors. Its visuals are clear enough: Facebook is all about people. You have pictures all over the place,[12] which are clearly sized up to provide us with spatial clues.

On your profile page, your picture on the left is the largest one. This is your place. If you go to the public home page, you still have your picture on the left, so you know it is still about you, but now its size matches those of pictures in the time line. This is a public space, and you are like everyone else. Friends and friends of friends commenting on posts appearing in the timeline have even smaller pictures. In that conversation, they are only guests (Figure 4.6).

Social networks that speak the language of place succeed. The rest linger or crumble: space exists independently of man, but place cannot. Place requires involvement, and in turn a sense of place is essential to our well-being. This calls once again upon the idea of context. Information architecture has always considered context as one of the key elements of design, but has usually intended context as the cumulative project constraints that weighed on the design process.

In pervasive information architecture, context is personal, social, existential context, connected tightly to the concepts of place and place-making, and spans channels. This in turn implies that place-making in pervasive information architectures has to work along two different axes: one internal and one external.

These are different. Internal place-making works in-channel, one artifact at a time, and it is aimed primarily at building the desired sense of place within the limitations or specific characteristics of that given channel.[13] External place-making strives instead to create spatial familiarity, comfort, and continuity across all the channels and artifacts that are part of the ubiquitous ecology: in a way, its reach is vaster and more profound–as it pervades the process–but less articulate and specific.

---

[12] Or, as Belgian information architect Peter Van Dijck put it, "And the faces! It's full of faces."
[13] We discuss how the specific nature of channels can impact the way we need to design and implement a sense of place in the case studies.

FIGURE 4.6
Place-making in Facebook through relative picture sizes.

LESSONS LEARNED

Know

- Space and place are different concepts
  Physical, objective, impersonal, stable—the former; psychological, subjective, experiential, dynamic, hodological, in one word, existential— the latter. Place is what we design in information space.

- Place is layered
  Place includes a relational layer of archetypes such as enclosure, vicinity, continuity, time; an emotional layer of feelings and sensations associated with the place; a behavioral layer of interactions and movements—either physical or semantical—inside the place itself
- Place-making has nothing to do with technology or the wow factor
  Place-making does not rely on technological breakthroughs but on the understanding of basic cognitive and psychological mechanisms that guide how we experience the world through our embodied self
- Context is more than a project's settings and constraints
  Context in pervasive processes is spatial and dynamic. It changes with the actors, the environment, the location, and the time.

Do

- Build place not space
  Allow for resilient way-finding: paths, edges, nodes, landmarks, and districts are dynamic, subjective experiences that can translate to semantic information spaces. Even here the shortest distance between two places is defined in accordance with the principle of least effort
- Make people feel at home
  Design for people, not users; design processes and stories, not products; design for concrete, situated interactions where people feel they can relate to the context
- Deploy both internal and external place-making
  Build a sense of place in-channel and across channels. Internal place-making adds to the character and sense of belonging of a single artifact in the ecology, whereas external place-making adds to a feeling of recollection and continuity across all artifacts.

CASE STUDIES
The Written Library

I cannot explain clearly what happened, but as we left the tower room, the order of the rooms became more confused. Some had two doorways, others three. All had one window each, even those we entered from a windowed room, thinking we were heading toward the interior of the Aedificium. Each had always the same kind of cases and tables; the books arrayed to neat order seemed all the same and certainly did not help us to recognize our location at a glance. We tried to orient ourselves by the scrolls. Once we crossed a room in which was written "In diebus illis," "In those days," and after some roaming we thought we had come back to it. But we remembered that the door opposite the window led

FIGURE 4.7
The Aedificium, the library of the abbey in Jean-Jacques Annaud's *The Name of the Rose.*

into a room whose scroll said "Primogenitus mortuorum," "The firstborn of the dead," whereas now we came upon another that again said "Apocalypsis Iesu Christi," though it was not the heptagonal room from which we had set out. This fact convinced us that sometimes the scrolls repeated the same words in different rooms. We found two rooms with "Apocalypsis" one after the other, and, immediately following them, one with "Cecidit de coelo stella magna," "A great star fell from the heavens."

(Eco 2006, pp. 194–195).

This is how, many years after the facts, Adso of Melk recounts his first bewildering visit to the magnificent but labyrinthine library of the abbey where the events narrated in *The Name of the Rose* took place (Figure 4.7). Written in 1980 by Umberto Eco, the book is many things to many different readers: a detective story, an intellectual game of deception, and an intriguing, if difficult, historical rendition of medieval Italy, as well as the final nonviolent expression of Eco's desire to "kill a monk."[14] It's easy to see how the novel is a "story of labyrinths, and not only of spatial labyrinths" with numerous **connections** to Jorge Luis Borges and his work.

> **Eco and Borges** - Way up to the point that we have a blind Spaniard, the venerable Jorge of Borges, as the real mastermind in charge of the library. Borges is a master of labyrinths, and as Eco himself pointed out that was simply due homage, as "library plus blind man can only equal Borges."

Adso recalls how he and his master and mentor brother William of Baskerville enter the library, find an immense treasure of manuscripts, and, as they wander through the rooms, get lost. They finally find their way out of the library again only by sheer chance when they are almost on the verge of giving up. But how did they end up there in the first place?

---

[14] As Eco himself described the original inception of the idea behind the novel.

In the book, the library is a large, almost windowless structure occupying the upper part of the Aedificium and only accessible during the day through a guarded door in the scriptorium, the hall where copyists work, below. Access is strictly controlled: books are dispensed for study or copy only upon permission, as the library contains both texts pious and heretic, and it's the librarian's privilege to say which is which. At sunset all doors are locked down and the library becomes inaccessible (or almost inaccessible).

But one monk seemingly either committed suicide or has been murdered in the abbey of recent, and brother William, who was an inquisitor once, is requested to investigate by the abbot. William has reasons to believe that the crimes may be connected to the library: nonetheless, he is repeatedly refused access. As the story unfolds, William and Adso break the rules and enter the Aedificium secretly and in the deep of night via a concealed passageway that leads from the cemetery to the kitchens and then to the scriptorium; as we have seen, their first visit almost ends up in disaster.

Finding a way to navigating the maze becomes paramount, and that is where it gets interesting for us. William considers using a compass and some elaborate methods involving markings on the walls, but then settles upon a powerful thought: they need to find "from the outside, a way of describing the Aedificium as it is inside." In the failing light of their third day at the abbey they take a walk around the building and note the shape of the walls, the position of the windows, and the number of the towers. Then Williams tells Adso to "try to draw a plan of how the library might look from above." Adso draws what we would call a sketchy blueprint of the library floors, and as the rooms materialize in front of him (he still does not understand fully what William is asking), he cries out "but now we know everything!" William comments that they know much, but not enough: they do not know where the openings are, if there is any logic to their position, and how the books have been distributed.

Describe the inside from the outside

They look at the crude map, and William solves the riddle. He first understands that those long sentences on the walls play some role. They are often repeated, which is unusual and unnecessary, as if those who built the library used them for some purpose. Then, suddenly, it's all clear: they are looking at, or imagining, a giant figured poem, "a cross or fish" as Adso says, where the initial letters of the scrolls make up words and possibly sentences: the library is a book and has an index.

A cross or fish

They visit the library again armed with their map and find out that their scheme works. Indeed the letters can be read into meaningful sequences, and these sequences in turn paint a map:

> In short, not to bore the reader with the chronicle of our deciphering, when we later perfected the map definitively we were convinced

that the library was truly laid out and arranged according to the image of the terraqueous orb. To the north we found ANGLIA and GERMANI, which along the west wall were connected by GALLIA, which turned then, at the extreme west, into HIBERNIA, and toward the south wall ROMA (paradise of Latin classics!) and YSPANIA. Then to the south came the LEONES and AEGYPTUS, which to the east became IUDAEA and FONS ADAE. Between east and north, along the wall, ACAIA, a good synecdoche, as William expressed it, to indicate Greece, and in those four rooms there was, finally, a great hoard of poets and philosophers of pagan antiquity.

(Eco 2006, p. 360).

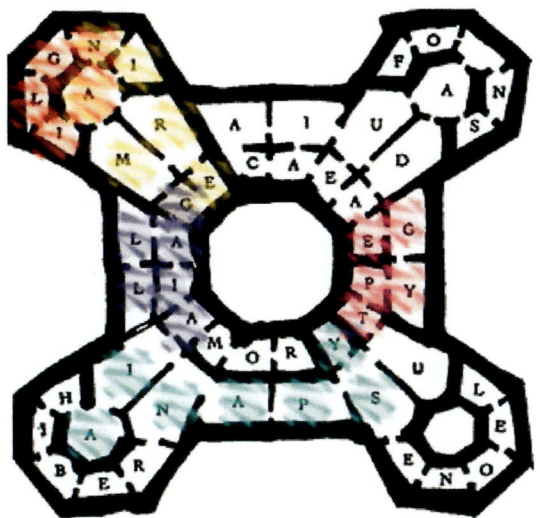

FIGURE 4.8

The map of the library from the book *The Name of the Rose*. Colors represent different regions of the world.

A map of the known world

Anglia is England, and Germani(a) the land of the Germans; Gallia is France, Yspania Spain, and Fons Adae the Paradise on Earth, in the East. This is all that is necessary to know to understand how the books are shelved, how to find and refind them, and how to move around in the library: the library is, in fact, **a map of the world** (Figure 4.8), and the books have been arranged so that Latin manuscripts are in Rome, and French authors are in one of the rooms of Gallia. Not only that, but someone looking for codices written in Anglia and moving from Aegyptus would know that they have to travel through Yspania and Gallia to get there.

The library and its devices are one single way-finding tool, but there is a twist. Even though the library in the book is not an incredibly complex maze,[15] someone rigged the rules to make it appear so. This library, contrary to your ordinary community library, has a darker secondary purpose: some of the books it contains should be lost and forgotten forever. One of these rooms is apparently inaccessible, and the library *has* to be difficult to navigate to the uninitiated or to the nosy and unreasonably curious; it's not by chance that in the casual distribution of the openings and passages, in the partial darkness, in the layout of the shelves, tables, and closets, all rooms just seem to be the same room.

But that's also the reason why some superstructure is necessary: to help those who know. The library is a complex, layered artifact (Figure 4.9) that shows how way-finding can be manipulated and distorted. Once conceived as a clever, commonly shared mnemonic device to help the monks move around

---

[15] There are only 56 rooms in total and they are all on the same floor. No stairways, no ups and downs, and no D&D-like traps or pits to confuse unwanted visitors.

FIGURE 4.9
Layered environments in Backseat playground, an experimental augmented reality game. Game elements are superimposed on the landscape as the car travels on. *Source: Interactive Institute, Stockholm.*

quickly and purposefully, the scroll and map system is now an instrument of power in the hands of those who want the library to preserve its secrets. That is also why a real, random labyrinth, a D&D dungeon, simply wouldn't work: it would hinder everyone.

What is interesting is that even though this conceptual labyrinth is imposed on the physical maze representing the world, these two do not entirely over-lap. William and Adso find out that the books are classified and shelved on the basis of a crystalline logic, **an underlying taxonomy**, that interprets and amends mundane "mistakes" when necessary:

An underlying taxonomy

> HIBERNIA, if we come from the blind room back into the heptagonal, which, like all the others, has the letter A for Apocalypse. So there are the works of the authors of Ultima Thule, and also the grammarians and rhetoricians, because the men who arranged the library thought that a grammarian should remain with the Hibernian grammarians, even if he came from Toulouse. It is a criterion. You see? We are beginning to understand something.

(Eco 2006, p. 352).

In a way, as grammar was eminently a Hibernian subject, all grammarians are or should have been Hibernia born: those who were born, say, in Egypt had their birthplace adjusted, and their books shelved accordingly. There is no such thing as a map of the empire: besides being a damning game of the goose in its arrhythmic sequence of openings, closures, and writings, the library is an interpretation of the world, a physical representation of an **ontology**,

A physical representation of an ontology

not the world itself: William's main accomplishment in the book is the uncovering of this arbitrary **mapping mechanism.**[16]

The Art and Craft of Being There

In the end, the murders are solved by chance, a most important book is lost, and William's mission fails. Andrea was decidedly luckier in Cambridge: nobody tried to kill him and he even found his way to the Sedgwick Museum. But at this point, you might wonder what we are trying to get to. A rotated street map in Cambridge and a fictional map in a book: so what, say you. Bear with us a little more, as we introduce one final element: the fact that one of the most intriguing things to happen to *The Name of the Rose* in the transition from book to movie was that the map of the library itself vanished through thin air.

The movie *The Name of the Rose*, directed by Jean-Jacques Annaud in 1986 and featuring Sean Connery as William and Christian Slater as Adso, is primarily a solid medieval whodunit with some interesting visuals but without a map. The layout of the library on screen bears no

The Library Map - William and Adso have mapped the library: they know that there has to be a central room in the South-East tower, the Finis Africae, a room William believes contains a book central to the murders, but they cannot find any doors or openings. "Secretum finis Africae manus supra idolum age primum et septimum de quatuor.": the key to the Finis Africae is that "the hand over the idol works on the first and the seventh of the four". When they finally solve the riddle and enter the secret room through a walled mirror - the idolum, from the Greek eidolon, image - it's midnight on their sixth day at the abbey. As Adele Haft says in her essay Maps, Mazes, and Monsters, an invisible door conceals an apparently inaccessible room. We cannot but notice that access to what lies beyond Africa, the monsters and lions of medieval maps and the evil books of this story, is only granted to those who walk through the looking glass. Which is an interesting thought to have in a conversation about being in cyberspace, being there and elsewhere.

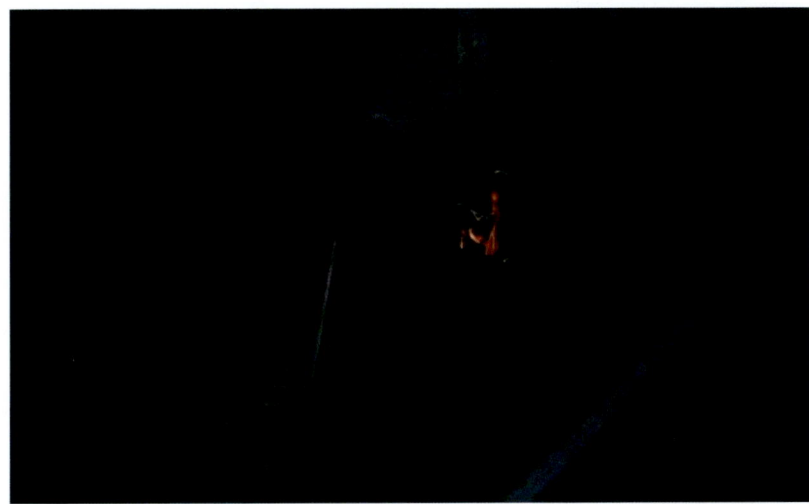

FIGURE 4.10
The library of the abbey in the movie.

---

[16] We guarantee that Clay Shirky would be proud of him.

FIGURE 4.11
The library as it was
rendered in the movie.
Photo: Anna-Maria C Sviatko
*Source: Theshoppingsherpa.
blogspot.com.*

resemblance at all to its literary counterpart. Because the movie is based squarely on the book, this prompts an interesting, if apparently trivial, question: why?

If you think about it and scrap the idea that Annaud just suffered heatstroke or got a little too enthusiastic on some local wine, there could be a simple answer: because the library in the movie needs to have the look and feel of a **labyrinth** to effectively disorient viewers (Figure 4.10). This prompts a second question: ok, but isn't the one in the book a maze? Wasn't that enough? Well, the answer is yes, and no. The written word and the visual representation work differently and use different languages. Representing these spatial labyrinths in such a way that the audience gets lost or feels they could get lost can be daunting, which is one of the reasons why Annaud and his crew decided to transmogrify the library into a physical, Escheresque labyrinth[17] with plenty of stairs going up and down and trompe l'oeil in numbers to deceive the unwary (Figure 4.11). In this process, everything concerning a map, scrolls, or the order of the world got edited out of the plot.

Now here's a fundamental question: why do we not usually get lost in movies, to the point that we need to be shepherded visually into a mannerist view of a labyrinth, something that screams "you will get lost here" at the top of its lungs? After all, we are not there. If it's enough to turn a street map 90 degrees right to confound us, shouldn't we get lost all of the time?

We do not, of course, unless the director wants to trick or deceive us.[18] The main reason being that we have a common, shared visual and semantic language for cinema, and that in force of that we are able to build a coherent, if imprecise and totally fictional, mental map of the space we see on the screen. It has not always been like that though: a brief look at the history of filmmaking shows how this common understanding has been long in the making. Early movies were conceived as pure filmed theater. Action

> **Eco and Labyrinths** - In one of his other many books on semiology, Eco provides a categorization of labyrinths and offers a differentiation between the terms *labyrinth* and *maze*. Labyrinths basically follow three models: classical, which is usually a spiral, mannerist, and rhyzome. Interestingly enough, the library in the novel is a manneristic maze of forking paths, and the rhyzome, or net, is a key element in the conceptual structure of the novel. The library reflects a world of forked paths, but the narrative, the events depicted in the novel, already prefigure a modern world of connections that the maze cannot hold inside anymore.

---

[17] Interestingly enough, this is the way it is also described in Manguel and Guadalupi's *Dictionary of Imaginary Places*.

[18] One of the best examples being Jodie Foster/Clarice Sterling ringing the door of serial killer Ted Levine/Jame Gumb in Jonathan Demme's *Silence of the Lambs*. Up to that point, the cuts and storyline have us thinking that the FBI is going to ring that door while Clarice is off on a side assignment somewhere safe.

on stage was recreated accurately in a one-room, flat, unarticulated space. There was no other space outside the current frame of vision, and the camera was positioned to present the audience with a theater-like experience.

In most scenes of E. S. Porter's *The Great Train Robbery* (Figure 4.12), shot in 1903, the audience and filmic space are aligned perfectly to mimic a stage representation, and actors enter and exit the scene from doors or openings on the left and right of the screen. The camera is perfectly still and pointed straight center. With all due consideration, this was not the effect of technological constraints: plenty of movies at the time tried what we would call special effects by means of movement and unconventional stage-tricks. But here there was a story. Rather, it was a semantic constraint. How to make the audience be there? What will they understand? Will they be able to follow?

It took years to develop a language that was capable of successfully conveying to the audience more than the simple one-room space of the theatrical stage (Figure 4.13).[19] What is interesting to us is that this had already been done. Some 20 years before Porter's masterpiece, English critic Theodore Child introduced his *Harper's New Monthly Magazine* readers to a new wave of French painters with these words:

FIGURE 4.12
E. S. Porter, *The Great Train Robbery*, 1903.

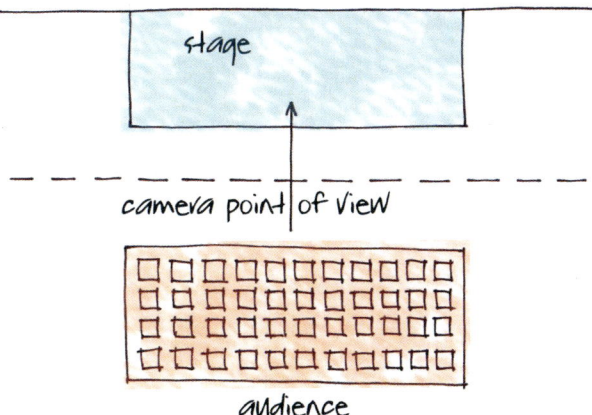

FIGURE 4.13
Early movie settings as theatrical scenes.

> Another marked peculiarity . . . is the truncated composition, the placing in the foreground of the picture of fragments of figures and objects, half a ballet-girl, for example, or the hind-quarters of a dog sliced off from the rest of his body. . . . It is the artist's means of showing clearly what his intentions are. . . . The composition is certainly strange, but it has a definite aim: it concentrates attention on the very parts where the painter wished it to fall. . . . There is thought and purpose in this apparent oddness.

---

[19] One of the most interesting elements was the development of ways to represent and understand deictic gaze so that audiences could follow the line of sight out of a scene and into another and connect them in a mental map. See Persson (2003)

FIGURE 4.14

E. Degas, *Musicians in the Orchestra*, 1872. *Source: Wikipedia.*

Child was commenting on the works of the Impressionists. He was especially keen on dissecting the works of Edgar Degas who, albeit being somewhat unnecessarily infatuated with washerwomen and ballet girls, he thought showed clear mastery of the craft. What Mr. Child was saying is that Degas, Monet, and Pissarro, among others, were painting scenes in which, we would say now, there was some unusual camera work and some place-making tricks were being employed. Take a look at *Musicians in the Orchestra* (Figure 4.14); this is no great train robbery, no flat screen in front of us comfortably seated in the audience. Degas has us right there where the action is. Look at the stage: can you see it in full? No, you cannot. The reason is that we are actually peering from over the shoulders of the musicians, part of the orchestra or right behind it. We are there, we are inside the painting.

So why weren't movie stage designers and directors simply picking up from there? After all, the Impressionists had such an impact on the visual arts that we cannot simply plead ignorance. The answer, at least for what little part is of concern to us, is plain: the similarities with theater were too great to be ignored from both filmmakers and the audience. They allowed moviegoers to take in this incredibly new experience of moving pictures in multiple locations while relying on a well-known frame of reference. Understanding cuts, using camera movements, and hinting at multiple spaces outside of the camera view–in other words, fully translating place from one medium to the other–could not be done without a specific, mature visual and semantic vocabulary that was not there at the beginning, but was there when Annaud directed his adaptation of *The Name of the Rose*. The library has become a visual labyrinth so we can get lost and be there.

RESOURCES

Articles

Chazan, M. (2004). Locating Gesture: Leroi-Gourhan among the Cyborgs. http://www.semioticon .com/virtuals/Locating%20Gesture.pdf.

Child, T. (1887). A Note on Impressionist Painting. *Harper's New Monthly Magazine.*

Dourish, P., & Chalmers, M. (1994). Running Out of Space: Models of Information Navigation. Short paper presented at HCI '94, Glasgow. http://www.dcs.gla.ac.uk/~matthew/papers/hci94.pdf.

Egenter, N. (1992). Otto Friedrich Bollnow's Anthropological Concept of Space. In *Proceedings of the 5th International Congress of the International Association for the Semiotics of Space*. Berlin, June 29–31. http://home.worldcom.ch/~negenter/012BollnowE1.html.

Haft, A. J. (1995). Maps, Mazes, and Monsters: The Iconography of the Library in Umberto Eco's *The Name of the Rose*. *Studies in Iconography*, 14, 9–50. Available online in a slightly modified form at http://www.themodernword.com/eco/eco_papers_haft.html.

Shirky, C. (2005). Ontology Is Overrated. *Clay Shirky's Writings About the Internet*. http://www.shirky.com/writings/ontology_overrated.html.

Simanek, D. E. (2006). The Flat Earth. Early Ideas about the Shape of the Earth. http://www.lhup.edu/ dsimanek/flat/flateart.htm.

Books

Bollnow, O. F. (1963). *Mensch und Raum* (Man and Space). Kohlhammer.

Calori, C. (2007). *Signage and Wayfinding Design*. John Wiley and Sons.

Dumézil, G. (1996). *Archaic Roman Religion*. Johns Hopkins University Press.

Falkheimer, J., & Jansson, A. (Eds.). (2006). *Geographies of Communication*. Nordicom.

Lawson, B. (2001). *Language of Space*. Architectural Press.

Lefebvre, M. (Ed.). (2006). *Landscape and Film*. Routledge.

Leroi-Ghouran, A. (1993). *Gesture and Speech*. MIT Press.

Manguel, A., & Guadalupi, G. (1999). *The Dictionary of Imaginary Places*. Harcourt.

Mitchell, W. J. (1995). *City of Bits*. MIT Press.

Monaco, J. (1977). *How to Read a Film*. Oxford University Press.

Munro, A. J., Höök, K., & Benyon, D. (1999). *Social Navigation of Information Space*. Springer.

Persson, P. (2003). *Understanding Cinema*. Cambridge University Press.

Piccaluga, G. (1974). *Terminus: i segni di confine nella religione romana (Boundary Signs in Roman Religion)*. Edizioni dell'Ateneo.

Tuan, Y. (1974). *Topophilia: A Study of Environmental Perception, Attitudes, and Value*. Prentice Hall.

Movies

Annaud, J. (1986). *The Name of the Rose*.

Greenaway, P. (1978). *A Walk through H*.

Porter, E. S. (1903). *The Great Train Robbery*. Available at American Memory from the Library of Congress. http://memory.loc.gov/ammem/index.html.

Consistency

FIGURE 5.1

ANDREA LEARNS SOMETHING FROM GAIA

Gaia is 7 years old and has a certain predilection for plushes. Animal plushes. It's late summer of 2007, and we are in her room in our Swedish home. This means that the current zoo on the shelves is a reduced, cut-down version of the larger one she takes care of in Italy.

I ask her: why do I always see these animals in small groups, and never all together? She gives me a look that says "Oh my, what a silly question, daddy," but she sets the core posse of plushes in front of me on the striped rug by her bed and tells me this tale.

Some of these animals are best friends, like the elephant, the donkey, and the seal. So they stay together all of the time. The two dogs stay by themselves, since they hunt down the others. Last week I had to put them to sleep (she uses the word *sedate*) more than once while they were setting up plush barbecues.

The rabbit is friends with these other animals (she points to the elephant-donkey-seal group), but comes from the same place of the sheep here (a farm? Gaia doesn't tell). So, he kind of stays in between, but only if the dogs aren't around; otherwise he hides in his hole.

The skunk and the penguin spend most of the day together since they are the same color. And the penguin now is teaching the skunk to swim, among other things.

Those two tiny tiny white seals are the same size, so they stay together as well.

The smaller elephant and the platypus are the same age and they love to play together, but I'm not sure they want to do that all of the time as the platypus has a beak just like the penguin, so sometimes they discuss beaky things just between them. (I ask what about the elephant, then.) The elephant has no beak, he has a trunk, you see, and he's not really interested in beaky things: if the platypus is not around, he sits there on his own.

A CHINESE ENCYCLOPEDIA

These ambiguities, redundancies and deficiencies remind us of those which doctor Franz Kuhn attributes to a certain Chinese encyclopedia entitled "Celestial Empire of Benevolent Knowledge." In its remote pages it is written that the animals are divided into: (a) belonging to the emperor, (b) embalmed, (c) tame, (d) sucking pigs, (e) sirens, (f) fabulous, (g) stray dogs, (h) included in the present classification, (i) frenzied, (j) innumerable, (k) drawn with a very fine camelhair brush, (l) et cetera, (m) having just broken the water pitcher, (n) that from a long way off look like flies.

Borges and Wilkins

Consistency - The capability of a pervasive information architecture model to suit the purposes, the contexts and the people it is designed for (internal consistency); and to maintain the same logic along the different media, environments and times in which it acts (external consistency).

This is the most famous and quoted paragraph from **Jorge Luis Borges's** *The Analytical Language of John Wilkins*, a brief essay in his *Selected Non-fictions*. John Wilkins was an English clergyman, a scholar, and certainly an ambitious thinker who, in 1668, published a book titled *An Essay towards a Real Character and a Philosophical Language*. Here he outlined his idea for a new analytical language to be used to describe the whole of human knowledge (Figure 5.2). As Borges (1972) reports

[Wilkins] divided the universe in forty categories or classes, these being further subdivided into differences, which was then subdivided into species. He assigned to each class a monosyllable of two letters; to each difference, a consonant; to each species, a vowel. For example: de, which means an element; deb, the first of the elements, fire; deba, a part of the element fire, a flame.

While criticizing the effective value of the four-level table, which is the base of the language in overly academic tones, and before moving on to discuss the efforts of the Bibliographic Institute of Brussels, Borges almost inadvertently ("These ambiguities, redundancies and deficiencies remind us . . .") introduces his fictional ancient Chinese list, which supposedly details how animals are to be classified. He is not the source though: he says he just reports what Franz Kuhn, a well-known German-born translator and scholar of Chinese literature, says. In this list, animals are divided according to incongruous principles: there are those that are the property of the emperor, those that look like flies from a distance, those that are fabulous (but we have a special category for mermaids), those that are tame, those that are finely painted, and those simply not listed there.

This list is weird, yes. It is fantastic, yes. It is certainly not the way you'd organize your zoo. Then again, doesn't it sound familiar? Imagine that in other less remote pages it is written that animals are divided into (a) friends, (b) hunters, (c) coming from the same place, (d) having similar sizes, (e) painted in the same colors, (f) with a beak. Does that ring a bell now? We are quite certain Borges wasn't thinking of plushes, at least those plushes, and even though we are pretty sure Gaia would love that Chinese list as much as we do, wild and imaginative as it is, we cannot ignore how that **taxonomy** is an astoundingly good example of what we should call an inconsistent classification scheme.

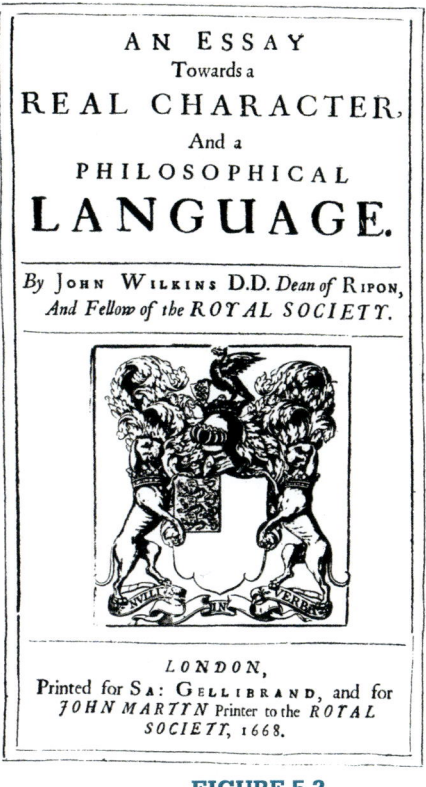

FIGURE 5.2
The 1668 frontispiece of John Wilkins' *An Essay towards Real Character and a Philosophical Language.* Source: Wikipedia.

Dear old Carl von Linné would definitely be upset about this, albeit in a very dignified Swedish way, and rearrange all of them quickly and properly in his Tree of Life (Figure 5.3). Or almost all of them. There, fixed it for ya, pumpkin. Now go play somewhere else. But who was Carl von Linné?

ONE TREE, SOME FLOWERS, AND A SWEDE

Carl Linné, or Carolus Linneus as his name was usually spelled in its Latinized form at the time, was born in 1707 in Småland, a region in central Sweden. He was a peculiar character, with a somewhat troubled start in life, and he most certainly developed through the years the same sort of self-confidence we attributed earlier to Mr. Wilkins, possibly to a larger degree. His personal motto in his mature years was "Deus creavit, Linnaeus disposuit," which modestly translates to "God created, Linné organized." On his account, and no offense to Mr. Wilkins, it must be said that Linné actually accomplished something in his lifetime, and this something, sprung out of his love for botany and plants, still bears fruits today: we call it the Linnean classification system.

Taxonomy - It refers to both (a) the discipline studying the classification criteria of a given set of items and (b) a specific classification system of knowledge organization. The term itself comes from the Greek and is a compound of *taxis*, meaning order, and *nomos*, meaning science. If we leave the discipline there on the shelf for a little while and we follow (b), we can say that taxonomies (plural) are classification systems where items are structured in a hierarchical tree or, more precisely, a set of classes departing from a main class usually called the root of the tree. Graphically represented, they usually resemble a tree turned upside down so that the root is the topmost element. Taxonomies are obtained by splitting a general usually complex concept, idea, or artifact in concepts, the classes, which are progressively more and more specific. All members in a class are marked by the same subset of shared features. The number of shared features decreases running along the branches of the tree and increases running up to the beginning of the tree; this means that upper classes possess all the features of the lower classes, but not vice versa. Taxonomies allow for a greater degree of precision in the classification process and support known-item seeking strategies, when users already know what they are looking for, very well. In recent years, a complementary term has been that of *folksonomies*—user-created, collaborative classification systems built by simple aggregation of tags, labels associated with content.

Linné and the *Systema Naturae*

Linné's initial catalogs just included plants: only later he extended them to include animals and minerals. It goes without saying that, even though Darwin and his *Origin of Species* had not stirred up the troublesome waters of evolution yet, he managed to get controversial.

For the joy of his contemporaries, he decided to classify plants based on their sexual characteristics: classes based on the number, length, and features of stamens, with orders based on pistils. He described flowers in such lyrical and sensual tones that he managed to attract a good number of rebukes mostly based on prudery alone. As Kennedy Warne (2007) reports, when Linné "described polyandrous flowers as having 'twenty males or more in the same bed as the female,' this was too much." Critics accused "bluebells and lilies and onions"–and good old Carl who uncovered the facts of course–of immorality, and pointed out how this was some "loathsome harlotry" which did not fit into the beauty of Creation.

Linné was not to be deflected though, and in hindsight it seems just natural that when he moved to animals he decided to have humans and apes as different genera in the same order, initially Anthropomorpha and then Primates. Close cousins, so to speak, which was a most scandalous idea. Even when he had to face theological allegations of impiety from both the Church of Sweden and the Church of Rome which had issues with his views, he maintained he was simply reporting the obvious, arguing in letters that he could not find one single generic difference between humans and simians in natural history, and that was the end of it. Linné worked on his classification system most of his adult life. In its first edition, in 1735, the *Systema Naturae* was a scarce 11 pages long and it was of course written in Latin, which was the scientific language of the time. By the time it reached its 10th edition, the one still considered for scientific purposes today, it was more than 800 pages, and it finally settled for around 3000 pages in its 13th and final edition.

At the root of the complete Linnean system are the three kingdoms (regna, singular regnum) he called *Regnum Animale*, *Regnum Vegetabile*, and *Regnum Lapideum*, that is, animals, plants, and minerals. The whole system was built

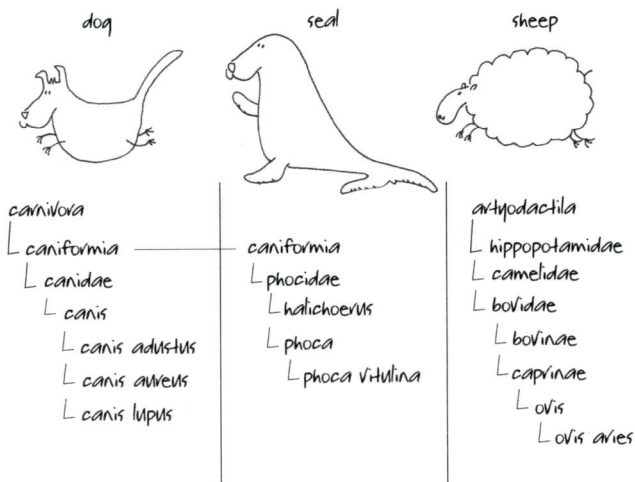

FIGURE 5.3
Gaia's plush animals placed correctly in the Tree of Life.

like a single hierarchy, a tree with branches, in which these three kingdoms were divided into classes, and these into orders, then genera, and species, and the grouping and splitting were based on shared physical characteristics.

Today most of his writing has pure historical value, as much as changed since he devised the Tree of Life: thanks to DNA sequencing, for example, we know now that hippopotamuses are related much more closely to whales[1] than to either horses or pigs, as Linné thought. Nonetheless, his work embodied the very idea of good, sound, Aristotelian classification and made it common knowledge: a system that limits or entirely avoids heterogeneous criteria, which is based on structured hierarchies relying on the use of one single sectioning principle, what is called in the High Speech a *fundamentum divisionis*, the conceptual basis of how we split things. One hell of a legacy, Mr. Wilkins, and incidentally one that completely obscured d'Alembert's and Diderot's almost contemporary and totally new device of arbitrarily structuring human knowledge according to the alphabet in the Encyclopédie.

RIGHT OR WRONG, MY CLASSIFICATION

Linné seems to suggest that if some animals belong to the Emperor, some animals are embalmed, and some animals are innumerable, there is simply no classification possible. And he for sure wouldn't be interested. Let Borges play his tricks and charms, but we know better: those animals should be distributed

---

[1] See *Science Daily* (2005).

neatly along the many branches of some revised version of the Tree of Life or in a similarly conceived taxonomy, where the Emperor and his whims and fantasies have no place: line Gaia's plush animals up and have them classified orderly along a top-down hierarchy: Animalia (kingdom), Chordata (phylum), Vertebrata (subphylum), and Mammalia (class). Except for the beaky penguin, whose class would be Aves (birds). And then have each in turn neatly categorized in its own order, genera, and species. Now, isn't that simpler?

That's the message of *The Analytical Language of John Wilkins*. Jorge Luis Borges, a librarian at heart, is adamant: the Chinese list and Mr. Wilkins's four tables on the universal language seem to make no sense at all. But this does not make classification impossible, quite the contrary. Sometimes the unexpected is better handled by the ingenious and imaginative.

PART FISH, PART BIRD, PART MAMMAL

It looks like a car that was built on a Friday. They used the parts they had left to put it together.

(Batzer)[2]

You might have noticed that Gaia had a platypus in her zoo (Figure 5.4). The girl loves cuddly animals, which means that as soon as one was found one day in some airport, she had to have it. It was brought home, was given a name, became friends with the elephant and the penguin, found its place, and that was it.

The platypus

Linné was some 20 years in his grave by the time the first report of an **Australian "water mole"** hit Europe around the end of the 1790s, but he would have definitely thrown a tantrum if he had still been around. The problem is, kids love it, but a platypus is a serious issue: not only does it look exactly like some god found some spare animal parts, grafted them together, and was pleased with the results, but its biology is baffling. When a first specimen consisting of a dried skin was shipped to England by Governor John Hunter, the beak, fur, tail, and webbed feet with a venomous spur were enough to excite and outrage taxonomists and zoologists alike. But when later on reports from Australia and direct observation hinted at the possibility that this puzzling creature laid eggs and milked its offspring, then the real fights began.

Fish, bird, mammal, reptile: the platypus was all of these and none of these, something that does not bode well if your reference is the ordered beauty of

---

[2] Batzer quoted in Bosveld (2009). Mark Batzer is a professor at Louisiana State University. Thanks to the work of Batzer and his team on DNA sequencing, we now know that the platypus genome is an extraordinary mixture of the reptilian, the avian, and the mammalian.

Linnean taxonomy and all you want is quick, painless classification. And it sure took more than 80 years to move the poor thing from "hoax" and "insult to God" to a real animal with its place in the Tree: "Monotremes oviparous, ovum meroblastic," "the monotreme lays eggs, and these are reptile-like," an excited W. H. Caldwell–a Cambridge zoologist who had been sent to Australia to investigate the animal–wired back to England in 1884, finally settling the issue. Even then, it was only in the purposely created, make-it-fit, order of the Monotremata, egg-laying mammals,[3] where it still sits today with the echidna as its sole companion. Trees have this way of falling just short of it at times.

FIGURE 5.4
A platypus. Photo: U. Djasim.
Source: Flickr.

CLASSIFICATION WANTS TO BE USED

It's not just about being neat, though. As Godzilla movies keep reminding us, size matters as well. It matters if you are an enormous mutated iguana on a spree, and it matters if you work with information. If the Web has taught us some lesson, it's that we suffer no scarcity of data nor will we in the future: information is not going away easily. And when dealing with large, complex domains or data sets in the googolplex range of sizes, the rigorous use of one single **sectioning** principle is often impossible or is a recipe for later problems: if anything, remember the platypus.

Sectioning complex domains

It may very well be that the primary goal of any classification is to provide scientific (in a very loose sense of the word) organization for a given knowledge domain, but classification systems also need to address day-to-day cataloging issues and empirical information management practicalities. Information has to be used: classifications need to adapt and be useful. Even if a classification system starts out with a maximum of scientific intent, as time and use go on, its original (scientific) architecture molds into a more empirical structure, one that the passage of time, culture, and context of practice imposes.

This dichotomy, rigorous vs. empirical, is basically irresolvable: Geoffrey C. Bowker and Susan L. Star offer some insight on how to deal with it in *Sorting Things Out*, published in 1999. Neatly balanced between the sociology of knowledge and technology, history and information sciences, this beautiful, fundamental book argues with plenty of supporting case studies that every

---

[3] The general implications of these shortcomings have always been a concern of information architecture. Rosenfeld and Morville mention the "darned" platypus in the Polar Bear book in Chapter 1, "Defining Information Architecture," and, sure, these authors used it as a prop to explain the implications of Aristotelian classification in various conferences.

classification system is linked deeply and inextricably to the social layers that produce it, and vice versa: all social systems deeply reflect themselves in the classification systems they produce. The book explores the *fault line* between folk classification, the one we constantly operate as we live our lives, and scientific classification, "a fracture that is constantly being redefined and changing its nature as the plate of lived experience is subducted under the crust of scientific knowledge" (Bowker & Star 1999, p. 67). One of the best and shortest examples deserves to be reported in full (Bowker & Star 1999, p. 38):

> Howard Becker relates a delightful anecdote concerning his classification by an airline. A relative working for one of the airlines told him how desk clerks handle customer complaints. The strategy is first to try to solve the problem. If the customer remains unsatisfied and becomes very angry in the process, the clerk dubs him or her "an irate." The clerk then calls the supervisor, "I have an irate on the line," shorthand for the category of an irritated passenger. One day Becker was having a difficult interaction with the same airline. He called the airline desk, and in a calm tone of voice, said, "Hello, my name is Howard Becker and I'm an irate. Can you help me with this ticket?" The clerk began to sputter, "How did you know that word?" Becker had succeeded in unearthing a little of the hidden classificatory apparatus behind the scenes at the airline. He notes that the interaction after this speeded up and went particularly smoothly.

THE ORDER OF THINGS

Foucault and the *Order of Things*

We are not entirely done with Jorge Luis Borges and our friend John Wilkins yet: that Chinese encyclopedia unexpectedly inspired a most important philosophy books, **Michael Foucault**'s *The Order of Things*, which has consequences for us. Foucault himself recounts in the preface that

> this book first arose out of a passage in Borges, out of the laughter that shattered, as I read the passage, all the familiar landmarks of my thought—our thought, the thought that bears the stamp of our age and our geography—breaking up all the ordered surfaces and all the planes with which we are accustomed to tame the wild profusion of existing things, and continuing long afterwards to disturb and threaten with collapse our age-old distinction between the Same and the Other. This passage quotes a "certain Chinese encyclopedia." . . . In the wonderment of this taxonomy, the thing we apprehend in one great leap, the thing that, by means of the fable, is demonstrated as the exotic charm of another system of thought, is the limitation of our own, the stark impossibility of thinking that.

It's easy to see what Foucault finds so exhilarating in this list that couples animals, mythological creatures, and what belongs to the Emperor: the sense of estrangement, the opening up to possibility. Foucault does not simply dismiss the encyclopedia as the incoherent blabbering or intellectual practical joke it was possibly meant to be, but rather wonders what makes it possible, what holds it together, why reading it we willfully accept to enter a caravanserai of wonders. And he asks but one single philosophical question, which has incredibly deep implications for information architecture and the design of human–information interaction: when we classify, when we say that a dog and a cat are more far apart than two Chihuahua dogs even if they are both trained or embalmed, where does this judgment stem from? What is this consistency and coherence we strive for if it is not predetermined by logical chaining or based on anything tangible?

Think of a painting, and imagine a city landscape. You might have a view of Venice in your mind, such as those by Giovanni Canal, the Canaletto, painted in the 18th century (Figure 5.5). You might have an Impressionist view of the streets of Paris, the ones that Gustave Caillebotte or Claude Monet loved to capture in the 19th century.

Or you might think of Frank Miller's Gotham city in the *Dark Knight Returns* graphic novels. Wherever your inspiration comes from, you probably see buildings, people, streets, sidewalks and waterways, yards and shops, and the flow of daily life at the time it was fixed on canvas or paper.

FIGURE 5.5

G. A. Canal, Canaletto, *The Stonemason's Yard* (1726–1730). Oil on canvas. *Source: Wikipedia.*

FIGURE 5.6

Masaccio, *The Tribute Money*, Cappella Brancacci, Santa Maria del Carmine, Florence. *Source: Wikipedia.*

Now we show you another painting. Close your eyes. Now open them and see mountains and a medieval city as Masaccio envisioned them in *The Tribute's Money*, a fresco in the Cappella Brancacci in Florence, Italy (Figure 5.6). The narrative does not follow the canon we expect from a painting, that of a frozen frame in time, but it's more like a movie, with events happening in temporal

STEFANO BUSSOLON—CLASSIFICATION AND COGNITION

"Everyone can make mistakes," said the hedgehog to the brush. It might not be a joke, but being able to identify an animal as a member of your own species, and as male or female, is a basic survival skill. Any animal incapable of such categorization would end up extinct in a flash. However, certain animals—and plants—make excellent and systematic use of the cognitive challenges inherent in categorization for survival. Some butterflies seem to have eyes painted on their wings: their purpose is to fool predators into thinking they might be looking at dangerous, larger birds and move the butterfly from (a) possible prey, hunt, and eat to (b) potential danger, stay away.

Creating categories is also a base mechanism for forming concepts: conceiving the idea of a bed involves understanding what a bed is and what it is not. A number of theories have been formulated to explain how we categorize.

It is generally accepted that the *classic theory* of categorization was developed by Aristotle and has remained almost unchanged all the way to Wittgenstein and the ethnographic and psychological studies of the 1960s and 1970s (Smith & Medin 1981). Classic theory is formalizable, efficient, can explain a vast array of phenomena, and is based on a limited and well-defined number of clear assumptions. According to it, a concept is characterized by a set of defining attributes, which are semantically necessary and sufficient for considering it an instance of an idea. Basically, to categorize means to verify that something has all the necessary characteristics. Classic information architecture assumes this view implicitly, that for each domain there is one and only one way to classify information. The task of the architect is simply to uncover this rule, codify it, and apply it.

The cognitive sciences have challenged this view: for example, Barsalou (2003) explicitly argues that there is always

STEFANO BUSSOLON—CLASSIFICATION AND COGNITION—CONT'D

more than one way to classify a domain and that these produce different categorizations depending on the context, on the circumstances, and on the goals. As a result, a number of different theories have been proposed to overcome this limitation.

Prototype theory suggests that categorization is a comparison between what we want to classify and prototypes of the various categories. When we stumble upon something we do not know how to handle, we assign it to the category whose prototype is the most similar. Smith and Medin (1981) say that prototype theory relies on the representation of a concept as the summary description of a whole class; the representation of a concept cannot be expressed by means of a list of necessary and sufficient conditions.

Prototype theory assumes that the classification will represent content and not boundaries, that the prototype embodies and represents the characteristics of a class, and that a graded approach is in place. Decision making is based on the similarity between the object or the concept that needs to be classified and the mental image of the prototype.

Because prototype theory has shortcomings and presents anomalies, other models have been proposed. Among them are, for example, exemplar theory, situated simulation theory, and decision-bound theory.

Currently, after some decades have passed, there seems to be no agreement on a prevailing model. Or better, many agree that each one of these different models captures some important aspects of categorization, and probably the focus of researchers should be on understanding when to apply them, when one is better suited than another (Ashby & Maddox 2005; Medin & Rips 2005).

Empirical research work seems to support this idea, and recent research in the neurosciences and in neuropsychology suggests that there are at least two different concurrent mechanisms at work when we classify: one based on rules and one based on boundaries. According to Ashby and Spiering (2004) and Ashby and Maddox (2005), individuals use different types of categorizations for different tasks, which in turn activate different areas of the brain. A series of card sorting tests conducted in 2007 confirmed this hypothesis and the fact that these two mechanisms can be used simultaneously and integrated.

Stefano Bussolon is an Italian psychologist, information architect, and usability specialist. He holds a Ph.D. in cognitive sciences and is currently a contract professor in data analysis at the University of Trento, Italy. He specializes in the sociocognitive aspects of interaction and participative information architecture and mostly works with card sorting, exploratory data analysis, and clustering algorithms.

sequence across the painted space. Peter appears in the same scene, left, center, and right, and performs different actions. Moreover, the temporal sequence does not flow from left to right as we are used to, but moves starts at the center, moves left, and then finally right. Because there is no *real* common perspective, spaces, buildings, and people are difficult to place: one character has his feet between the mountains and the city, clearly two different nonadjacent spaces. It's hardly a painting we understand without specific training today, and we have no doubt Masaccio would find Batman pretty difficult to understand as well, let alone enjoy Miller's cinematographic cuts and storyline.

So let's be even more radical: close your eyes and think of one of Escher's woodcuts (Figure 5.7). Here every perspective, every wall, window, door, or column is the result of decisions. It's still a city, but whether that is the floor or the ceiling, whether that is hollow or solid, what is up and what is down is something that viewers have to decide on their own. Similarly, *The Order of Things* argues that similarities and differences, the essence of categorization and classification, are but the outcome of a given set of preliminary, either individual or

FIGURE 5.7
M. C. Escher, Complex and Concave. *Source: Wikipedia.*

Classification is arbitrary

The Dyirbal language

collective, choices. Whatever these might be, they are not abstract and natural at all. They are specific, personal, cultural, and socially meaningful.

Classification is not far from some cognitive Escheresque pattern (Figure 5.8): in order to go beyond the initial bedazzlement and displacement these perspectives cause, we have to take a decision, call that surface "floor" and that one "ceiling." Because such sense making is totally **arbitrary** or wholly subjective, we can close our eyes and easily subvert it to obtain a completely different, if not a totally opposite, picture or we can read a list as a list, even if it contains wildly unrelated items or bewildering animal categories. Foucault has turned the table upside down: now we can take this thing one step further and meet Mr. George Lakoff and the Dyirbal language.

FOUCAULT AND LAKOFF

Borges, of course, deals with the fantastic. These not only are not natural human categories—they could not be natural human categories. But part of what makes this passage art, rather than mere fantasy, is that it comes close to the impression a Western reader gets when reading descriptions of non-western languages and cultures.

(Lakoff 1987, p. 92)

Traditional **Dyirbal** is an aboriginal language of Australia whose classification system is embedded right into the language. In the words of Lakoff, "whenever a Dyirbal speaker uses a noun in a sentence, the noun must be preceded by a

FIGURE 5.8
A scene from Christopher Nolan's *Inception* (2010) featuring a Penrose staircase in the style of M. C. Escher's *Ascending and Descending* (1960). *Source: Screenshot from the movie.*

variant of one of four words: bayi, balan, balam, bala. These words classify all objects in the Dyirbal universe." These four base categories shape the Dyirbal language and also provide the title to his book: women, fire, and a number of apparently unrelated dangerous things are all part of the same category, *Balan*.

Lakoff maintains that elements in categories are correlated using analogical models such as **metaphors and metonymy** rather than classic Aristotelian links (Figure 5.9).

This is a major departure from the traditional model of classification, the one Linné used in his *Systema Naturae*. In Aristotelian taxonomies, categories or *classes* are abstract containers, and every item is either in or out. Possessing certain properties makes items belong to a certain class and hence clearly help identify the class itself. This is *in re* the nature of things as they are. For example, all fish and only fish have gills all throughout their life: this is what constitutes the boundary and what makes them belong to their class.[4] Aristotelian systems imply well-defined boundaries, equal importance for all items in a class, and no active role to be acknowledged to those building the classification. This is why the platypus was such a huge issue in the early 1800s: it *had* to be a hoax, as it was not possible that something factual could be fish, bird, and mammal at the same time. After all, Linné was simply describing what classification God had imprinted into the world, and fish, birds, and mammals are meant to occupy different places in the order of things.

And Dyirbal, imbued as it is with the language's culture and vision? Well, Dyirbal is an incoherent primitive tool. Gaia's groupings are a kid's fallacies,

Metaphor and metonymy - Metaphor describes something as being the same to something else in some way. This way, implicit and explicit attributes from the second subject enhance the description of the first. A canonical example is "the ship plowed through the sea," where the image of plowing a field is applied to the motion of a ship in the water. Metaphor comes from the Greek *metaphérein*, meaning "to transfer."

A metonymy is the substitution of a term with a second one associated by either logic or contiguity: "the sails crossed the sea," where sails is a metonymy for the whole ship, or "reading Borges," where Borges is a metonymy for his books and writings. Again the word originates from the Greek, where *metonymía* means "change of name."

Both figures involve the substitution of one term for another, but where one is based on similarity, the other is based on contiguity. Lakoff and Johnson (1980, p. 3) consider metaphor and metonymy fundamental mechanisms in our mental processes.

FIGURE 5.9

Prototypical elements in the four Dyirbal classes. Squares represent central elements (prototypes); circles represent nonprototypical elements; and connectors represent the analogical connections (chains) that link prototype to nonprototype elements. *Source: Lakoff (1987, p. 103).*

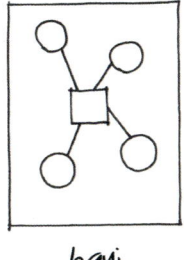
bayi

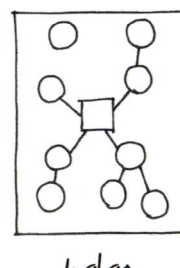
balan

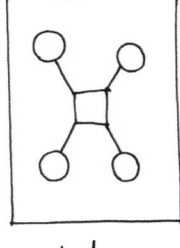
balam

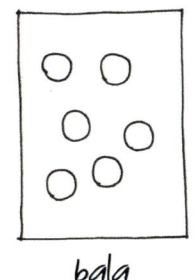
bala

---

[4] As opposed to amphibians such as frogs, which have gills during their tadpole stage but not as adults.

the Chinese encyclopedia is an eccentric fantasy. And we agree, they might be. But only if seen against the background of a totally different system, such as a different culture, or the abstract rigorousness of Linnean classification. When in context, they all make perfect sense.

THE CHAIR AND THE RUG

You could legitimately think that Dyirbal is an extreme example, so let's try a little experiment and read the following words:

messer	tisch	gabel
land	name	stadt
mädchen	tag	nacht
kind	mond	sonne
meer	ozean	see

Now, what if we tell you that there is this language, German, spoken by people who live in a country with a little sea in the north and plenty of mountains in the south, where the masculine gender includes such diverse things as a table (tisch), the moon (mond), a train (tag), and the ocean (ozean)? Where the feminine gender spans forks (gabel), cities (stadt), the night (nacht), and the sun (sonne)? And where there is a third neuter gender, making up for a good 20% of the words you find in the dictionary, which include, among others, country (land) and girl (mädchen)?[5] Does Dyirbal still look so extreme?

The relationship between language and classification runs deep and is a frozen unsteady lake we won't venture across. What we can say, and what is important in Lakoff's presentation of the Dyirbal language, is that there lies a good example of a whole different classification model, relying on whole different logic, and it works.

Eleanor Rosch and prototype theory

Eleanor **Rosch** elaborated her **theory of prototypes** in the 1970s on the basis of field work she conducted among the Dani in Papua New Guinea. Now a professor of psychology at the University of California Berkeley, Rosch published a momentous paper titled *Natural Categories* in 1973. There she argued that everyday classification, folk classification, or wild classification as it is sometimes called, relies less on abstract definitions of categories than on a comparison of the given object or experience with what is deemed to be the object or experience best representing its category. Rosch had observed that the Dani were capable of classifying objects of colors they linguistically did not differentiate, as they possessed no words for English hues and chromatics and maintained

---

[5] German is not the only language to present gender structures and differences. Spanish, French, Russian, and Italian, for example, all present similar scenarios.

only a distinction between what is "light, bright" and what is "dark, cool." Rosch concluded that people in different cultures tend to categorize objects by using prototypes, elements that are the initial stimulus associated with that specific category, and that these prototypes may vary. In a subsequent paper published in 1975, *Cognitive Representation of Semantic Categories*, she asked American college students to rank a number of items according to their being representative of the category *furniture*. She found out that chair, sofa, and table topped the list, while shelf or rug scored very low: they were, in a way, less capable of conveying the idea or essence of what furniture is.

Rosch refined her theory through the years and came to elegantly define prototypes as the most central element in a category. This, of course, means different things in different contexts: a chair may be the prototype for furniture in North America, while it surely won't be such for the nomadic Sami of northern Scandinavia. As a rule, in prototypical systems

- elements of a class do not share similar properties;
- some elements are more *central* and represent the whole set or class much better than other elements, such as a sparrow for birds or a chair for furniture. These are called *prototypes*: opposition and mutual exclusion, if any, rely on them as they identify the category;
- central elements are related to peripheral members by chaining and analog mechanisms, such as similitude, metaphor, and metonymy;
- there is or there might be a class *other*, which comprises whatever elements do not belong to the other classes. Such class neither has prototypes nor uses chaining, as it is the case with the residual category *Bala* in the Dyirbal language.

In Bill Moggridge's *Designing Interactions*, Brenda Laurel, a famous researcher, entrepreneur, and pioneer in the field of human–computer interaction, author of the fundamental *Computers as Theatre*, reports an interesting anecdote. She joined Interval Research in 1992 with the goal of building electronic games that could appeal to girls. At the beginning, that meant a lot of research to identify and frame their gender-related problem space: literature runs, but also work in the field. In the course of these investigations they interviewed roughly a thousand preteen boys and girls, and they unexpectedly found out that some characteristics are more important than others in defining the basic gender nature of an artifact (Moggridge 2007, p. 355):

> We made a pink furry truck, and learned that pinkness overrides truckness. We did a diary with bullet holes in it, and found out it is still a diary and a boy won't use it.

More recently, Umberto Eco (1997) connected this phenomenon to grading and salience, elements Rosch identified in her research, in his book *Kant and the Platypus*. Artifacts have **primary characteristics** and **secondary characteristics,**

Primary and secondary characteristics

the former being core and necessary, with the latter being somewhat more peripheral and unessential. For example, you can easily have elephants without their tusks, but not without their trunks. Trunks embody a large part of their *elephantness*. Similarly, Laurel's elements of *truckness* were overridden by the color pink, which was considered by their teen sample to be a primary characteristic for *girly stuff* as much as *diaryness*, no matter how many macho-man bullet holes they drove through it.

CONSISTENCY IN PERVASIVE INFORMATION ARCHITECTURE

I have registered the arbitrarities of Wilkins, of the unknown (or false) Chinese encyclopaedia writer and of the Bibliographic Institute of Brussels; it is clear that there is no classification of the Universe not being arbitrary and full of conjectures.

(Borges 1972)

Whatever our stance on classification, classical, prototypical, empirical, or something else entirely, it is rather clear that the very ideas of coherence and consistency are grounded in the time and culture from which they originated, and there is no Grand Unified Classification in the Sky to resort to. Scientific systems become more and more a mixture of the empirical and the arbitrary as time passes, and tell the platypus about it. Consistency needs to be assessed in

DONNA SPENCER—BASIC LEVEL CATEGORIES

So what's this idea of a basic level category?

When we think about categories—about things and concepts—we think more often about some kinds of objects than others. For a hierarchy of things (and everything fits in some type of hierarchy) we actually think in the middle of the hierarchy. We neither think at the very broad level of a hierarchy nor do we think at the very detailed level. Continuing the furniture example, let's look at this hierarchy (Figure 5.10).

In this hierarchy, the basic level will usually be around the bed/chair/table/bookcase level. We don't think about furniture or about office chairs, but we do think about chairs.

Basic level categories are described as having some of the following characteristics:

- They are learned early
- They have a short name that is in frequent use. The name also feels like it is the "real" name for an object.
- You can often imagine the category with a simple visual representation (e.g., it's hard to imagine furniture but easy to imagine a chair)

DONNA SPENCER—BASIC LEVEL CATEGORIES—CONT'D

FIGURE 5.10

- There may be a representative action taken for the objects (e.g., chairs are for sitting in; there is no consistent action for "furniture")

One of the most important ideas behind basic level categories (and also applies to all types all categories) is that they are not absolute. You can't look at a hierarchy and choose the basic level. The basic level of a hierarchy depends on the person doing the thinking. The more people know about a subject, the more their "thinking level" becomes detailed. So a city dweller may think at the level of tree/bush/shrub where a country dweller may think of oak/maple/ash/eucalyptus.

So what does this all mean for real-life pervasive information architecture? My take:

- Basic level categories can be likened to topics or subjects—all the things we think about all day.

- Topics are natural categories for which to organize information around.
- When we are looking for information and answers, we use those topics to guide us, for example, we think "I want to know more about chairs," not "I want to know more about furniture."
- Topic-based or subject-based information architecture will often be much more useful than an audience, task-based or audience-based information architecture (you could reference this article rather than go into this: http://www.uxbooth.com/blog/classification-schemes-and-when-to-use-them/).
- You can often spot basic level ideas from user research—these are the topics that people talk about most often (e.g., for an intranet, people always talk about travel, HR, finance, social).
- When you are assembling a hierarchy, start with basic level categories/topics. This forms the core or middle of the set of ideas you are working with. Aggregate them into broader categories and break them down into detailed categories. You'll have more success with this than if you just try to start at the top of the hierarchy and break down things bit by bit.
- When you design a set of information, help people get to the topics/basic level categories quickly (these are great "quick links") and let them explore the content from there.

Donna Spencer is an Australian freelance user experience designer who specializes in large, messy Web sites and large, messy business applications. She has written three books on card sorting, Web writing, and information architecture. In her spare time she runs UX Australia, an annual user experience conference.

respect to a system's context, goals, users, and cultural climate that produced it in the first place and within which it lives.

In pervasive information architectures, consistency is layered: we have an **internal consistency**, related to the single artifact, collection, or organization system, and an **external consistency**, related to multiple connected artifacts or to systems linked together. While the traditional information architecture model deals mostly, if not exclusively, with internal consistency, for example, if

Internal and external consistency

all parts of a Web site work together, a pervasive model has to consider the shift from product design to ubiquitous ecologies design. Consistency also needs to be evaluated as it relates to the different media and environments that a single service or process spans.

Salience

As simplistic as it may sound, there is no right or wrong classification model to refer to, but only a certain degree of fittingness to a given task. This fittingness, this effectiveness, is coupled tightly with **salience**. Patrick Lambe (2007) reports and comments on an interesting case of salience in the taxonomy of solid cancers; these are currently classified by the parts of the body in which they originate, but

> oncologists now believe that a biochemical classification makes more sense, because similar biochemical mechanisms underlie cancers that can appear in various parts of the body, and understanding the mechanism can improve detection, prognosis and treatment. The salient organizing principle is no longer location, but mechanism. It's not that a classification by location is wrong, it's just not especially useful any more.

Salience is a key dimension in the way a given system can respond to issues of consistency. A salient system is robust enough to filter out noise and to grade its elements, becoming more consistent in the process. Classification systems with low salience have low consistency as well and are generally less capable, but salience cannot be measured or observed by ways of abstract Aristotelian principles: it is an empirical, context-aware, sense-making indicator.[6]

Incidentally, it may also be worth pointing out that if there is no right or wrong classification, probably there is no right or wrong theory of classification either: it is quite possible that different theories capture different aspects of the social and information complexity they want to represent and that we effectively employ different mechanisms in different contexts for different purposes.

Classification models have far-reaching ethical, political, and moral implications that impact directly on everyone's life, all the time. As individuals and as societies, we categorize and classify from the moment we wake up in the morning. For this reason, every classification system is but a mixture of the rational and the empirical, stirring prejudice, common sense, and scientific criteria into an uneasy mortar we then dry, use, and keep together with generous amounts of duct tape.

Winter White Russian hamsters—*Phodopus sungorus*, Linné would quickly add—are a species of hamster originating in the Siberian steppe, usually no more than 10–12 cm in length (Figure 5.11). They have a thick dark gray dorsal stripe and cutesy furry feet. Winter White Russians owe their name to the fact that their coat can turn white during the winter, and they are skilled diggers

---

[6] Bowker and Star (1999) discuss several examples where large-scale classifications evolved more robust system salience through the years.

and hazardous climbers. Luca's Russian hamster, Nebbiolo, is an explorer as well, and we wouldn't be surprised at all to find one of his kin for sale on eBay side by side with some plushes and a map of northern Russia.

It will be just like reading in some other obscure encyclopedia: in its remote pages it will be written that the animals are divided into (a) Russian hamsters; (b) fresh eggs; (c) plushes of different sizes; (d) that are depicted in fine inks on a map of Siberia; (e) Snoopies; (f) not in this list.

But then again, if more than one possible classification scheme exists for a given set, how do we get to choose one over the other? How do we evaluate internal or external consistency in any given classification scheme, if these cannot be evaluated abstractly, beforehand? Again, a few rules of thumb could be helpful. Here's a recap.

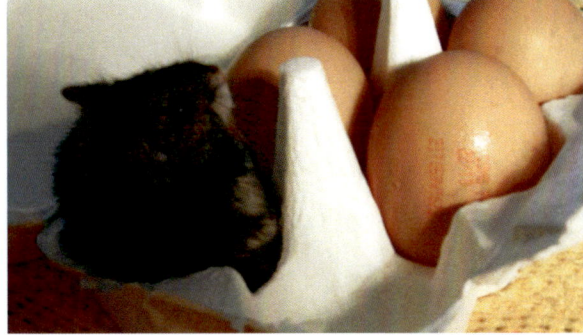

FIGURE 5.11
What is a Winter White Russian hamster doing inside an egg carton? Does a category "fresh eggs and Russian hamsters" even make any sense?

LESSONS LEARNED

Know

- Consistency is contextual
  Consistency needs to be assessed in respect to an empirical paradigm: its context, its goals, its users, the cultural climate that produced it in the first place and within which it lives. As simplistic as it may sound, there is no right or wrong, but only a certain degree of fittingness to a given task.
- Consistency spans the process
  In pervasive information architecture, consistency has two faces. One is internal and has to do with the general salience of the system; the other is external and relates any artifact within the ecology with the ecology as a whole.
- Consistency is miscellaneous
  Miscellaneous categories such as "other" do have a place in our view of the world, even if their use in an information architecture may prove to be problematic (in designing navigation, for example). This might identify elements that can be marginal in the design process, especially in the early stages.

Do

- Use a graded prototype approach
  Whatever the scheme, most likely there will be no ordered, clear-cut border between every two categories, but these will exist between central elements, the prototypes. Using these to represent classes helps users understand and choose.

- Follow an up-and-down model
  The classification process starts in the middle, from the prototypical categories, grouping them in supercategories and then splitting them in subordinate, more specific categories. Prototypicality works within a hierarchy as well so that middle-classes are usually more representative of the whole tree.
- Make base categories easily accessible
  Users should be able to reach base-level categories as quickly as possible. It's entirely possible that popular tags in folksonomies identify prototypes.

CASE STUDIES

A Taxonomy for Snoopy

A very interesting case study was made public in 2003 by Katherine Bertolucci. In *Happiness Is Taxonomy—A Classification for Snoopy* she recounts her experience in building the new catalog and related complete new taxonomy for Determined Productions, the largest Peanuts toys company on the market (Figure 5.12). Bertolucci was given a clear goal: the new classification had to offer an ordered framework to entice and stimulate new ideas in the team. This is interesting per se, as the new catalog had to be designed with a dynamic, innovative twist to it that served first and foremost employees. The primary target were not customers (Figure 5.13).

As it turned out, Bertolucci designed an incoherent classification scheme that worked swell. The main categories she finally settled upon were

Babies	Dolls	Housewares
Bed and bath	Electronics	Kitchen
Books	Fashion	Music
Christmas	Figurines	Office
Decorations	Garden	Plush

FIGURE 5.12

Teddy and Snoopy (detail).
Photo: G. Kat. *Source: Flickr.*

A few considerations. The categories follow different sectioning principles: by type (book, doll), by target (babies), and by use (kitchen, bath, office), but once more this inconsistency is a defect only if read out of context. If in abstract a plush is certainly some kind of doll, inside Determined Productions a plush was a high-selling item, while dolls were a minor sector with potential for growth: the separation was meant to prevent plushes from eating up dolls and to clearly highlight that these had some niche value to them, as they deserve a place in the list. Moreover, categories are ordered alphabetically along the

FIGURE 5.13

A+R online store sporting a main menu (on the left) split in two, actions and materials. Bertolucci went for another approach.

principle of most convenience so that important (company-wise) items come first. *Babies*, for example, had little use at the time but was seen as a strategic sector to be developed.

RESOURCES

Articles

Bertolucci, K. (2003). Happiness Is Taxonomy: Four Structures for Snoopy: Libraries' Method of Categorizing and Classification. *Information Outlook*, March. http://findarticles.com/p/articles/mi_m0FWE/is_3_7/ai_99011617.

Borges, J. L. (1972). The Analytical Language of John Wilkins. In Borges, J. L. *Other Inquisitions: 1937-1952*. University of Texas Press. (Available in a slightly different translation at. http://www.crockford.com/wrrrld/wilkins.html).

Kurowski, B. (2006). On the Order of Things. *Brandt.Kurowski.net*, October 29. http://web.archive.org/web/20071116055905/http://brandt.kurowski.net/2003/10/22/wilkins/.

Spencer, D. (2006a). Four Modes of Seeking Information and How to Design for Them. *Boxes and Arrows*, March 14. http://www.boxesandarrows.com/view/four_modes_of_seeking_information_and_how_to_design_for_them.

Spencer, D. (2006b). Lakoff's 'Women, Fire & Dangerous Things': What every Information Architect should know. In *Proceedings of the 7th Information Architecture Summit*. Vancouver, March 23–27. ASIS&T. http://iasummit.org/2006/.

Nutt, D., King, A. L., Saulsbiry, W., & Blakemore, C. (2007). Development of a Rational Scale to Assess the Harm of Drugs of Potential Misuse. *The Lancet, 369.* http://www.thelancet.com/journals/lancet/article/ PIIS0140673607604644/abstract.

Resmini, A., & Rosati, L. (2007). From Physical to Digital Environments (and Back): Towards a Cross-context Information Architecture. In *Proceedings of the 3rd European Information Architecture Summit.* Barcelona, September 21–22. http://www.euroia.org/2007/Programme. aspx.

Books

Bowker, G. C., & Star, S. L. (1999). *Sorting Things Out: Classification and Its Consequences.* MIT Press.

Foucault, M. (1994). *The Order of Things: An Archeology of the Human Sciences.* Vintage.

Kuhn, T. (1962). *The Structure of Scientific Revolutions.* University of Chicago Press.

Lakoff, G. (1987). *Women, Fire, and Dangerous Things: What Categories Reveal about the Mind.* University of Chicago Press.

Stephenson, N. (2003). *The Baroque Cycle.* William Morrow.

Video

Marconi, I. (2010). Definisci la cosa. (Defining the Damn Thing). *Contest for The 4th Italian Information Architecture Summit.* Pisa, May 7–8. http://vimeo.com/11338696.

CHAPTER 6

Resilience

FIGURE 6.1
Photo: Umbria Lovers.
Source: Flickr.

LOOKING FOR THAT SPECIAL WINE

Think of visiting an elegant wine shop when Christmas approaches. All around us is a fine scent of wood and spices, and the shop is crowded with people who are looking for a gift or for something they can take to a dinner. All staff members are rather busy. In front of the many shelves, a number of different customers scan the labels. Chances are, three of them have some "special" need or request, which in turn implies a different way of seeking, a different way of searching the shelves. Sure enough, all of them are looking for wine, and all of them have chosen this particular winery because they know it has one of the best equipped cellars in the area. Here they can find bottles that really cater to wine lovers. Let's take a closer look at what they are doing.

The first customer, the younger guy in a jacket, is head over heels with this girl he met a few months back. He is looking for a white wine with a character he associates with his girlfriend, perhaps with a fruity flavor, not too strong. The second customer, a seasoned businessman with a love for good cuisine but little knowledge of wines, is looking for something to go with the green pepper fillet he promised a couple of friends who are coming over to dinner. Another friend at the office told him he needs a robust red wine, aged in a barrel, and that's all. The third customer, the lady with that fashionable coat on the right, is the only one who has a clear idea of what she is in for: she absolutely wants a bottle of Donnafugata. They praised the wine on her favorite morning talk-show today, and she hastily wrote down the name. She probably did not even get it right, she knows that, but that's okay, this shop is one of the best around. Now she is curious and impatient and looks forward to a quiet evening with good food and a little wine.

Now, this is a gourmet winery, so care has been taken to arrange the wines on the shelves according to country of provenance first and region second. As in many retail shops, this is a monodimensional approach to classifying goods that produces a standard, easily accessible taxonomy—that is, if you are an expert. In our scenario, which is far from being exceptional, our three customers are not going to be able to find what they are looking for easily without asking someone from the staff. And this is going to be time-consuming at the least, as everyone is terribly busy.

As compelling as it is, as sound as it is, the geographical layout is not really helping anyone to have a better shopping experience. The lady in the coat is the only one who might stand a chance to find her bottle, if she could only remember that the wine she is looking for comes from Italy and, more precisely, from Sicily.

That winery is a place for experts: the kind of people who already know what they are looking for, those who employ what we call a **known-item seeking strategy**.[1]

HUMAN–INFORMATION INTERACTION

A synchronic society generates trillions of catalogable, searchable, trackable trajectories: patterns of design, manufacturing, distribution and recycling that are maintained in fine-grained detail. These are the microhistories of people with objects.

(Sterling 2005, p. 45)

---

[1] Seeking strategies are basic behavioral patterns we use when looking for information. Known-item is one of them and implies that the user knows what she is searching for and how to describe it. For an in-depth discussion in connection to information architecture, see Rosenfeld and Morville (2006) or Spencer (2006a).

Unfortunately, not everyone is an expert, and even experts are experts only in very specific areas: more often than not they will find themselves in situations where they, like the rest of us, are absolutely clueless. And not just that, as sometimes experts just do not want to play expert, but want to go with the flow or with the experience of others.

Our needs and wants deeply influence the way we look for things, our **seeking strategies**. This is true for every task we perform and every environment we find ourselves in when we search for products, information, services, or people, either online or in the street. Cognitive, cultural, and social models have a strong impact on behavior; as a result, we all are different individuals who browse and search differently because we all have different goals and possess different reference models.

Seeking strategies differ

If this was not enough, we modify our patterns according to context and time. We look for the same items in different ways at different times: if we are late for work, we hurry out of the house. If we cannot find that book we wanted to read on the bus, we squirrelly browse a few common places and get more and more frantic and imprecise as the seconds tick away. If we know that the bus is still more than half an hour away, though, we might proceed methodically and without sweating it that much. Maybe we even decide to pick up something else to read or go for some knitting. Context plays its part as well: imagine that book was a gift and you took it to the park for some reading and resting on the grass. The sun is up, and you nap for a while. When you wake up, the book is nowhere to be found. Your seeking strategies are going to be radically different and might involve a visit to the police precinct.

Two **characteristics** actively shape this process of human–information interaction and impact either positively or negatively:

Two characteristics shape the human–information interaction process

- ■ the capability (or incapability) of an information space to adapt itself to the needs of its users
- ■ the capability (or incapability) of an information space to support multiple information seeking strategies

This is the moral of the story of the Crowded Winery and the Three Unsatisfied Customers given earlier: from the perspective of information architecture, every information space measures up to these two crucial properties. Together, these are what we call **resilience**:

> *Resilience* - The capability of a pervasive information architecture model to shape and adapt itself to specific users, needs, and seeking strategies.

the fundamental capability or incapability of an information space to shape and adapt itself to different specific users, needs, targets, and seeking strategies.

Implementing or designing resilience in cross-channel user experiences requires a radical shift in the way we look at issues of information retrieval and information architecture, as smart, mobile devices and people need to be part of the picture to improve the system's behavior, to make it more resilient. This means that we are not dealing with the way a closed, mostly static system answers questions; we are dealing with a dynamic environment where what users do with available information changes or influences the current and future status of the system.

This is nothing new: Gary Marchionini underlined how the information retrieval community is very well aware of this necessary step forward in an article he wrote in 2004:

> there are increasing efforts in the IR research community to incorporate people into the retrieval problem. . . . Thus, the problem shifts from the system optimizing matching to put burden on the human information seeker to engage in an ongoing process. In such a user-centered paradigm, people have responsibilities and capabilities.

Spimes

If we want to get the visionary angle, we only have to resort to Bruce Sterling and his *Shaping Things* (2005, pp. 76–77), replete with the possibilities offered by synchronic items, **spimes**, objects capable of living in space and time that users coproduce and harvest as information wranglers:

> You first encounter the SPIME while searching on a Web site, as a virtual image. The image is likely a glamorous publicity photo, but it is also deep-linked to the genuine, three-dimensional computer-designed engineering specifications of the object—engineering tolerances, material specifications and so forth. Until you express your desire for this object, it does not exist. You buy a SPIME with a credit card, which is to say you legally guarantee that you want it. It therefore comes to be. Your account information is embedded in that transaction. The object is automatically integrated into your SPIME management inventory system. After the purchase, manufacture, and delivery of your SPIME, a link is established through customer relations management software, involving you in the future development of this object. This link, at a minimum, includes the full list of SPIME ingredients (basically, the object's material and energy flows), its unique ID code, its history of ownership, geographical tracking hardware and software to establish its position in space and time, various handy recipes for post-purchase customization, a public site for interaction and live views of the production change, and bluebook value. The SPIME is able to update itself in your database, and to inform you of required service calls, with appropriate links to service centers. At the end of its lifespan, the SPIME is deactivated, removed from your presence by specialists, entirely disassembled, and folded back into the manufacturing stream. The data it generated remains available for historical analysis by a wide variety of interested parties.

AN INTEGRATED MODEL OF INFORMATION SEEKING

Information seeking, a field related to information retrieval but keen on understanding how we search, has explored in depth the ways people look for information in different environments. It is an optimal starting point for a conversation that is trying to produce some ideas about how to better integrate people and information as coactors in a single, dynamic process.

However, the different theories of information behavior that have been proposed through the years have often considered the relationship between human beings and information from one point of view at a time, either focusing entirely on the social or the cognitive, the anthropological or the biological, the online experience or the offline, everyday experience.[2] Scholars pursuing a holistic, integrated approach have been scarce. Among them, Marcia Bates, now professor emeritus of Information Studies in the Graduate School of Education and Information Studies at the University of California, Los Angeles.

In her famous article "Toward an Integrated Model of Information Seeking and Searching" (2002), she proposes an **integrated**, non specific framework for seeking and searching behaviors as general human-information interactions.

An integrated model for information seeking and searching

Bates distinguishes four fundamental strategies for information seeking and harvesting, which can be composed in a bidimensional matrix (Table 6.1). The vertical axis measures the level of consciousness, and the horizontal axis measures the level of voluntariness:

- vertical axis (consciousness)
  directed: individuals can specify to some degree what they are looking for
  undirected: individuals cannot articulate their need; they expose themselves to information randomly
- horizontal axis (voluntariness)
  active: individuals acquire information actively
  passive: individuals absorb information passively (from family, friends, colleagues) and do not enact any active seeking behavior.

Table 6.1 Fundamental Strategies of Information Seeking (Bates 2002)

	Active	**Passive**
Directed	Searching	Monitoring
Undirected	Browsing	Being aware

---

[2] For a compendium of the various theories and models of information behavior, see Fisher and colleagues (2005).

The resulting matrix illustrates the four fundamental strategies we can adopt while engaged in information seeking behaviors.

1. *Searching*: directed and active information seeking. We are conscious we need a certain piece of information that we are able to articulate and we work actively to find it.
2. *Monitoring*: directed and passive information seeking. We are conscious we need a certain piece of information that we are able to articulate, but we do not search actively. This modality identifies a propensity to absorb pieces of information from the context without engaging in a direct search, which relates to serendipity.
3. *Browsing*: undirected and active information seeking. We have no specific interest or need or we cannot articulate it, but we acquire new information actively.
4. *Being aware* (or awareness): undirected and passive information seeking. We have no specific interest or need or we cannot articulate it, and we do not acquire information actively. We rather absorb it from the context.

As Bates herself explains, this means that "monitoring and directed searching are ways we find information that we know we need to know," whereas "browsing and being aware are ways we find information that we do not know we need to know."

The most interesting and surprising insights actually come by examining the relative importance these different strategies have in everyday life, not just online: Bates argues that about 80% of all our knowledge is absorbed passively and indirectly by simply *being aware*; about 14% by *monitoring*; 5% is sought actively by *browsing*; and only an exiguous 1% is acquired actively and directly by engaging in *searching* (Figure 6.2). Think about this for a second: it means that almost all of our knowledge (and to be precise a dazzling 94%) comes to us by means of **passive** information harvesting, just by "being conscious and sentient in our social context and physical environment" (Bates 2002).

Search is mostly passive

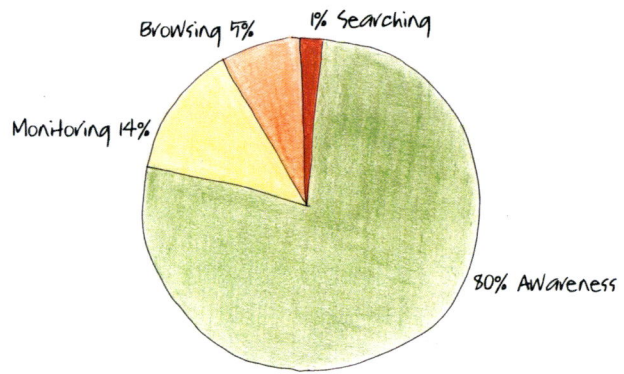

FIGURE 6.2

Percentage relevance of the different information seeking strategies in everyday life.

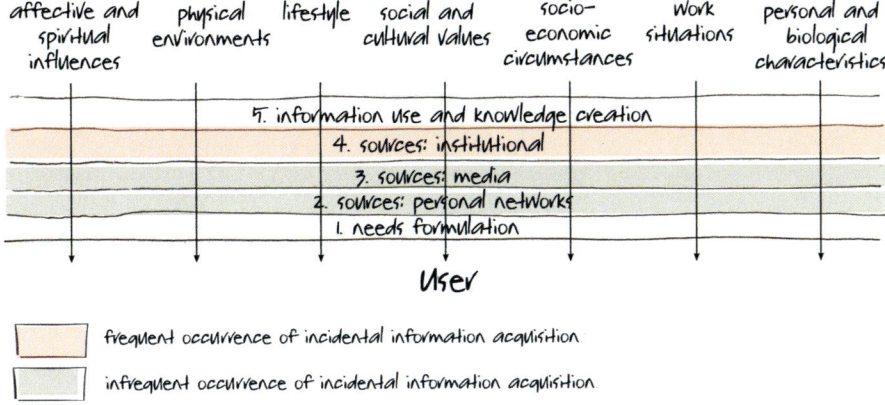

FIGURE 6.3

Kirsty Williamson's integrated approach to information seeking behavior.

Kirsty Williamson (2005), a long-time LIS researcher now at Charles Sturt University in Australia, has more recently developed Bates' seminal model further and has proposed an *ecological* approach to human–information behavior that, as much as Bates's, postulates close contiguity between cognitive and social layers. Williamson focuses on the relationships between information resources and their users, and, most importantly, does not stop at the boundaries of Web space but considers the broader picture of human–information interaction across channels (Figure 6.3).

Time for a little break. Fancy some bagpipes?

CLAUDIO GNOLI—FROM LIBRARIES TO KNOWLEDGE ORGANIZATION

Many techniques for the arrangement of information content have been developed in libraries. The subject of any document (a paper, a book, a disc, a movie, or a Web site) can be represented by means of a controlled vocabulary or translated into the code of a classification scheme in order to produce helpful sequences in both online catalogs and the library shelves. Schemes are designed in ways such that documents dealing with related topics have higher probabilities of being filed in close places.

Each user will have a particular topic in mind where to start her search (her own Umbral region, U). When looking there, she may find that some documents on the left and right of the starting point share some facets with it that may also be relevant for her search (the Penumbral regions, P). As she moves away from the focal topic, she will find a long tail of increasingly irrelevant topics (the Alien regions, A). Each user at each moment will have her own bell-shaped relevance curve: what S. R. Ranganathan has called the *APUPA pattern*. Despite trying to provide for user needs, classification schemes often

Continued

CLAUDIO GNOLI—FROM LIBRARIES TO KNOWLEDGE ORGANIZATION—CONT'D

FIGURE 6.4

A 16th-century fresco in a church in Stroppo (Turin, Italy) showing a variety of bagpipes currently unknown. Photo: C. Gnoli.

force them to think in a "learned" way, as their primary subdivisions are academic disciplines, like physics or sociology.

This may cause some problems with interdisciplinary documents or with documents intended for nonacademic readers, for example, children. The León Manifesto (http://www.iskoi.org/ilc/leon .htm) then claims that schemes should better serve interdisciplinary knowledge by adopting as information units the phenomena studied rather than the disciplines studying them.

Suppose that you are looking for information on the presence of bagpipes in the traditional culture of a certain geographic area. As bagpipes have fallen into disuse in many places where once they were widespread, very few documents in the discipline of musicology are available dealing directly with this topic. However, you could infer precious information on the presence of bagpipes at given times and places from nonstandard sources, such as travel notes of writers in past centuries, ancient police minutes reporting on unauthorized feasts that involved bagpipers, frescos with scenes where a bagpiper appears, specimens and recordings from the investigated region kept in faraway museums of artifacts and sound archives, puppet and crib collections, or organizations dealing with traditional dances.

All these potential sources are hardly indexed under the perspective of musicology. If the only information you can get is "Renaissance fresco with Nativity scene," how can you know that a detailed bagpipe is represented in it? Although many different information sources are now available through the Internet, their subject contents are not interoperable with each

CLAUDIO GNOLI—FROM LIBRARIES TO KNOWLEDGE ORGANIZATION—CONT'D

other if they are described simply as belonging to a given discipline (fine arts) or genre (sacred pictures) or form (frescoes). The only unit of information that is really universal is the phenomenon dealt with.

Thus, although library classifications offer a precious heritage of systems and techniques, in our cross-medial age we need to go a step further. We need to move from the arrangement of academic books on shelves to a more general notion of "knowledge organization," including indexing of information contents

in any form and medium. Since 1989, the International Society for Knowledge Organization (http://www.isko.org) has been exploring theoretical bases for such an approach. It is now up to systems developers to apply such updated conceptions to provide open and effective features.

Claudio Gnoli, Italian, is a librarian at the University of Pavia, Italy. He is active in various international organizations and the author of books and research papers in the field of knowledge organization. His Web site is http://mate.unipv.it/gnoli/.

THE PRINCIPLE OF LEAST EFFORT

Now that we've had a cup of coffee, let's sit down and consider Figure 6.2 again.

Clearly, we could say that there seems to be some conservative principle at work: most of the time, whatever floats the boat is okay with us. Or, the less energy we spend going in circles the happier we are, even to the point that we accept lower quality, compromises in services, or less reliable pieces of information if these are easier to obtain and use. We have all said at least once that something is "good enough," meaning that although not perfect, that something actually gets more or less the job done, now, without resorting to active research strategies. These necessarily involve further efforts, more time, and occasionally more skills, and we gladly avoid all of this if not strictly necessary. This fundamental human trait has been given many different names, and one of them is the **principle of least effort**. We mentioned it briefly in Chapter 4: now let's take a closer look, shall we?

The principle of least effort

We have long puzzled in this field over this human perverseness. Why do physicians not use the medical literature, rather than relying on the drug company salesman for information about a new drug? Why will our students not get up and walk a hundred meters to access a key journal article in the library? Well, put in the context presented here, we can see that throughout human history, most of the information a person needed came to him or her without requiring active efforts to acquire it. Picture the hunter-gatherer: Raised in a family group or clan, most learning came through interaction with one's mates and with the environment, that is, through being aware and monitoring . . . . Directed searching is further complicated by another factor in our modern lives. It has only been in the last 200 years or so that the amount of recorded information available has grown to such an extent that complex and sophisticated access mechanisms have had to be developed to enable access. So, people accustomed to mostly passive ways of learning new information not only have to search actively for the information, but also have to master a fair amount of ancillary skills and knowledge just to be able to search for the information, with no guarantee that that effort will actually lead to an answer.

Put in this way, I think we can see why the overwhelming propensity of most people is to invest as absolutely little effort into information seeking as they possibly can. It is only in moments of great urgency or great interest that they spontaneously begin investing seriously in acquiring the information skills needed to satisfy their needs.

(Bates 2002).

It's plain to see that the principle of least effort implies some kind of initial inertia. As humans, and throughout our evolution from cave dwellers to video game players, we have always been keen in gathering information passively from the context, be it the clan, tribe, family, or environment, as that provides the best almost-free meal we can get, so to speak. We have simply maintained such a propensity and passed it on to our children, resorting to active seeking only when passive seeking fails.

Satisficing

The principle of least effort is connected, among other things, to the concept of **satisficing**, a term coined by the economist Herbert Simon as a cross of *satisfying* and *sufficing*, meaning some degree of satisfaction obtained with minimal effort. Simon's point is that often we don't make optimal choices, we satisfice. This behavior has always had a special ring for us who design for the Web ever since we understood that competitors (or simply more interesting opportunities) are just one click away from any user. Steve Krug has widely used the concept in his usability-driven approach to the Web; in his highly successful book *Don't Make Me Think* he actually provides a very interesting overview of how satisficing impacts decisions in all aspects of life:

we don't choose the best option—we choose the first reasonable option, a strategy known as satisficing . . . . I'd observed this behavior for years, but its significance wasn't really clear to me until I read Gary Klein's book *Sources of Power: How People Make Decisions.* Klein has spent many years studying naturalistic decision making: how people like firefighters, pilots, chessmasters, and nuclear power plant operators make high-stakes decisions in real settings with time pressure, vague goals, limited information, and changing conditions. . . . They took the first reasonable plan that came to mind and did a quick mental test for problems. If they didn't find any, they had their plan of action.

(Krug 2005, p. 24).

INTEGRATING APPROACHES

While it surely provides a well-grounded context for satisficing, we believe that the principle of least effort has a larger, more crucial role: it neatly explains why the integrated approach underlying Bates's model is important, why human propensity to privilege passive modalities for information harvesting can be properly understood only by using a holistic view and, in turn, produce better results in a human–information interaction process.

Let's see: the biological and evolutionary points of view reveal it to be a heritage of our hunter–gatherer ancestors. Anthropology and sociology may contribute to explain why such behavior was a winning behavior in the first place, connecting it to the social sphere of any individual's life: because children gather the major part of their knowledge from their social habitat, this well-ingrained pattern persists even when they are adult and they keep resorting to **gathering** information from colleagues, peers, family, and a larger and more heterogeneous circle of friends. At the cognitive level, we may discover further interesting links and find out that the principle of least effort actually applies to philology, linguistics, and a number of other disciplines[3] as well and that in the end it might prove to be an intrinsic brain pattern, one of the ways we are wired.

Gathering information from the environment

Indeed, evolutionary biology recognizes a phenomenon called **exaptation** according to which a feature, originally developed for one specific purpose by a certain species, is then used for another, different purpose when pressure and demands from the environment change (Gould & Vrba 1982). Exaptation seems to explain both the passive (being aware and monitoring) and the active strategies of information seeking (browsing and searching). Passive modes have most probably been exapted from the natural propensity of our hunter–gather ancestors to learn by being immersed in their environment (tribe, clan, family); active modes of information seeking, however, have likely been exapted from a natural propensity to **sample and select,** which is typical of mating and foraging behavior. In other words, when it came to food and finding a partner, we were a little pickier than when dealing with news of novel ways you could sew a mammoth skin. Not surprisingly, we still are.

Exaptation - Browsing, berry picking, mingling, dating, shopping, nibbling, sightseeing, way-finding, channel surfing, Web surfing: all of these seem to be part of a larger, evolutionary chain that goes back to the sample and select techniques of our ancestors. *Exaptation* refers to the change some biological function or behavioral trait might undergo during evolution in order to better serve new or different needs. A similar concept can be found in linguistics, where the action of applying a new perspective to something in order to change its significance is called *reframing.* Reframing applies to a wide range of human experiences, from verbal communication to kinesthetic memories of past events, and either involves a semantic shift, where the meaning of something changes, or the superimposing of a new context of reference. For example, reframing is one of the base mechanisms in jokes and poetry.

---

[3] For more in-depth information on this, see Case (2005) in Fisher et al. (2005).

A FEW IMPLICATIONS

A brief recap of the major implications of what we have been saying so far seems necessary. Because of the way we use and consume information, classical information retrieval is not enough in pervasive information architecture. First, we now deal with human–information interaction across heterogeneous environments, and conversely we need a model that emphasizes the necessity of a holistic approach to the challenges at hand. Second, integrating people in the design problem space is necessary. Third, we need to consider human beings as the complex animals they are, taking into account a number of different facets—biological, anthropological, cognitive, and social—and different contexts—leisure time, workplace, family life, hobbies, friends, and so on.

> Surely, it is desirable to build our understanding of information seeking behavior on all the layers [below], not just some, whether upper layers or lower ones.
>
> Spiritual (religion, philosophy, quest for meaning)
>
> Aesthetic (arts and literature)
>
> Cognitive/conative/affective (psychology)
>
> Social and historical (social sciences)
>
> Anthropological (physical and cultural)
>
> Biological (genetics and ethology)
>
> Chemical, physical, geological, astronomical
>
> So the phrase, "integrated model," in the title has a dual meaning in this talk. I am attempting to 1) integrate our understanding of information seeking across the several levels, or layers, of human life, and 2) develop an integrated model of information seeking in relation to information searching. So where does information seeking come into this general context of integrated layers? First of all, let us consider information seeking with respect to all the information that comes to a human being during a lifetime, not just in those moments when a person actively seeks information.

(Bates 2002).

This final point connects directly to the *principle of least effort*. If we examine information seeking behaviors through the lens of this particular microscope, it does not seem unreasonable at all that roughly 80% of everything we know we simply gather by being aware, conscious, and sentient while living our lives in our social context and physical environment, as Bates maintains: this is what we have always been doing.

Hence, and this is where everything connects, an information system that wants to be resilient has to be designed in order to promote **serendipity** and to support user-modifiable **push** strategies, that is, passive or almost passive ways to acquire information. The dominance of passive strategies and the pressure exerted by the principle of least effort suggest that most of the time some informational needs are latent, or not completely conscious, and they become self-evident only when a relevant item is encountered. The possibility of altering the mechanism through time either explicitly or by means of self-improving filters and tools is then an integral part of the design.

Serendipity

> **Push** - Push describes a certain type of communications where transactions are initiated by the publisher or central server based on preferences expressed in advance by users or subscribers. Whenever new content is available, the publisher pushes this new information out to the users. E-mail and sensor network monitoring are both examples of push approaches.

Let's go back to our winery for a little moment: imagine that the lady who was looking for a bottle of Donnafugata is finally discussing a few options with a shop assistant. They are in front of the shelf labeled Sicily, talking and handling bottles. Another lady, waiting for some gift to be wrapped, catches snippets of their conversation, hears about how that wine complements the food she is going to have for dinner, becomes interested, and buys a bottle. This is an example of how being aware transforms a latent need (a certain dinner calls for a certain wine) into an explicit choice (buying that wine).

The sheer amount of data, information, and products available today brings along huge benefits in terms of possible choices, but it sure also makes the search and selection process infinitely more difficult, especially when the goal is not (well) known or cannot be articulated, as was the case again for at least two characters in our Crowded Winery Christmas Carol. That's the reason why the possibility of adapting, of becoming resilient by either contextualizing or personalizing choices or by providing ways for exploiting our natural propensity for normal approaches, or both, becomes critical.

RESILIENCE IN PERVASIVE INFORMATION ARCHITECTURE

Places are used as wax. They bear the layers of a writing that can be effaced and yet written over again, in a constant redrafting. Places are the site of a mnemonic palimpsest.

(Bruno 2007, p. 20).

Resilience is the capability of an information space to shape and adapt itself to different users, needs, and seeking strategies. A successful resilient environment implements its information architecture as a dynamic process where people are active players and an integral component of the design. This is extremely

FIGURE 6.5

Tracking and reusing patterns of use and communication in a dance project, part of Synchronous Objects. *Source: Synchronous.osu.edu.*

important in a pervasive process spanning different media or environments: by keeping track of the multiple stories narrated by the interactions between people and information, people and places, people and other people (Figure 6.5), and by allowing these in turn to shape or reshape the process, we build a stronger core identity that changes and flexes differently at the different layers and scales. By making these stories available asynchronously for later consumption and reuse, we help meaningful, purposeful patterns to be accessed effortlessly, with a minimum of awareness.

In his article "From Information Retrieval to Information Interaction" we quoted a few pages back, Marchionini seems to envision a similar scenario:

> Additionally, the system may save increasingly detailed traces of fleeting ephemeral states arising in online transactions. . . . Thus, our objects acquire histories, annotations, and linkages that may strongly influence retrieval and use. . . . What seems particularly promising are opportunities to discover new kinds of features within objects, and new kinds of relationships among objects that can be leveraged for retrieval purposes.
>
> (Marchionini 2004, p. 3).

You can imagine a down-to-earth, real-life situation if you think of your visits to a convenience store or a supermarket. If your interactions with the environment are preserved in some way, it suddenly becomes possible to think of ways that a customized mobile application, an intelligent barcode reader (either provided by the store and activated with a personal card or code, or residing on your smartphone), or a rabbit-sized djinn, for that matter, can help you

- refind paths
- customize your shopping and save time, money, or mileage
- receive meaningful, personalized push suggestions and correlations
- share your histories and profiles with family or friends

This is also what Giuliana Bruno, professor of Visual and Environmental Studies in the Department of Visual and Environmental Studies at Harvard University, hints at when she speaks of places as **mnemonic palimpsests** in her 2007 book *Public Intimacy*. Preserving a history of the flow and actions of people inside places transforms them effectively into a text, into architextures, emotional landscapes, and mobile maps where the environment is complemented by the interactions of people with them. In a way, in Bruno's vision people are like pens: they write the stories of their interactions with places inside those very places, and hence inevitably change them. As we said when introducing place-making, places have a spatial component and an existential, emotional, personal, and social part that stretches back and forth into the past and into the future. Exploiting these sediments, these narratives, help make them resilient.

Places as palimpsests

A similar view is expressed by Peter Merholz of the San Francisco–based UX firm Adaptive Path. To recompose tagging and user-generated content into a self-contained idea of design, Merholz (2009) introduces the idea of *desire lines*

FIGURE 6.6
Desire paths are often opened even to avoid a simple deviation from the least effort. *Source: Flickr.*

Desire paths in
information space

or *desire paths*, a concept borrowed from urban design: "trails worn into a landscape that demonstrate the paths people want to take, not those that were laid down by the designer" (Figure 6.6). Merholz uses spontaneous urban artifacts to explain how designers should be able, at least to a certain extent, to read and then implement in information space user behaviors that were not part of the original design. This can be taken a little bit further. We already established that digital and physical are not opposite sides of a coin but rather places we inhabit differently: what about seeking strategies? How do they manifest themselves? For example, what happens if we compare searching a typical Web site with searching a supermarket in the light of Marcia Bates's four modalities model (Table 6.2)?

Desire paths, palimpsests: these could be encoded and read across environments, like applications such as Layar or Wikitude, augmented reality applications for smartphones–which superimpose user-produced information on a view of the real world–seminally do. This is the landscape that the SENSEable City Laboratory at the Massachusetts Institute of Technology is busy painting. Real Time Copenhagen is an experiment that was carried out in 2008, which measured the "pulse of [the city's] Kulturnatten [the culture night] as it unfolds in real time." Both passive information, connected to mobile phone positioning, and active information, via volunteers tracked by means of their GPSs, are displayed on a map and connected with ongoing events (Figure 6.7). The city comes alive in its flows and social trajectories, in its hubs and bottlenecks. Everyone using the map to decide what to do next instantly becomes also an

Table 6.2 Correspondences between Information Seeking Strategies in Digital and Physical Contexts

Digital	Physical	Information Seeking Strategies (Bates)
Search	Specific places, objects, people having unique IDs or coordinates	Searching
A–Z index	Alphabetical list of items and related coordinates	Searching
Main and local navigation	Departments, aisles, shelves, and similar	Browsing
What's new	New items, hot topics, promotions, or highlights	Browsing
RSS, newsletters	Push alerts	Monitoring
Shortcuts	Custom paths for returning users or specific targets/needs	Monitoring/being aware
Social navigation	Popular items or paths	Being aware
Contextual navigation	Related items	Being aware

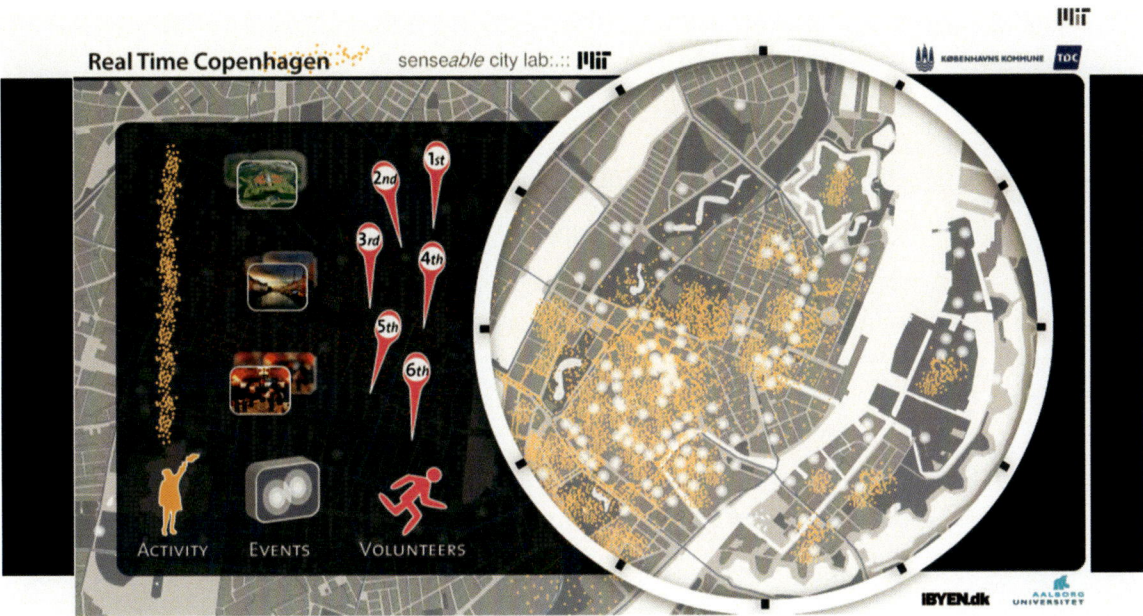

FIGURE 6.7
Real Time Copenhagen,
one of the seminal
applications built by MIT
(this one in collaboration
with the City of
Copenhagen and Aalborg
University, Denmark).
*Source: Senseable
.mit.edu.*

agent of change, adding his quantum of information to the pool and effectively reinforcing the descriptive and predictive capabilities of the system.

When we mix the physical and the digital, we give a voice to objects, places, and long-gone interactions and we make them talk to us: a few of the complex interactions between what they say, how we can record that, and what we can do with it are explored in Table 6.3.

Designing a resilient information space means conceiving an adaptive environment flexible enough to support different seeking strategies, directed and undirected, active and passive; providing it with enough push to inject a sufficient degree of serendipity; and making it capable to restructure itself according to the changing and heterogeneous interactions, actions, and needs of its users, considered as biological, cultural, and social beings. Making it capable of weaving stories:

> Imagine the sensory overload of a walk in the park. Every path
> shimmers with the flow of humanity. Every person drips with the
> scent of information: experience, opinion, karma, contacts. Every tree
> has a story: taxonomies and ontologies form bright lattices of logic.
> Desire lines flicker with unthinkable complexity in this consensual
> hallucination of space and nonspace, a delicious yet overwhelming
> sociosemantic experience.

> (Morville 2005, pp. 153–154).

Table 6.3 How, What, Why: How We Save Interactions, What This Provides Us, Why It Is Important

How	What	Why
Geotagging, unique IDs via RFId or similar code systems	Every item may be localized and is directly findable	Enables direct search for objects in the physical world Allows customized paths
Recording paths via smart cards, mobile devices, and similar Cross-referencing recorded paths	Usual or custom paths may be refound for personal or social use Amazon-like correlation strategies such as "if you like *x* maybe you also like *y*" or "people who saw this also saw"	Refinding frequently used paths allows optimization Custom paths can be shared Receiving suggestions in push mode
Tagging enabled by RFId or similar technologies and by mobile devices	Tagging and collaborative tagging for improved metadata on items	Used for both personal purposes (refinding items; creating wish lists . . .) and social purposes (receiving suggestions or discovering related items by people with similar profiles; sharing paths or lists . . .)
All of the above	All of the above	Monitoring interactions and flows help the corporate to improve the information architecture of the entire ecosystem

We hear you: this is way complex, dudes. Yes, but complexity does not stop at our door just because we would love it to. Complexity has to be embraced. And while it's true that complexity can span environments and channels in mostly unpredictable ways, handling it does not necessarily go hand in hand with "complicated," as biologist Eric Berlow (2010) points out correctly in an incredible short TED talk titled "How complexity leads to simplicity". In his presentation, Berlow shows how an overly "complex" diagram of the U.S. intervention in Afghanistan can be broken down visually in smaller, simpler, actionable steps. What is unpredictable and complex at project time zero, before all the thinking and the design start, might actually be assessable either later on in the process or by the simple adoption of visuals and network logic.[4]

We deal with this issue more in detail in the next chapter on reduction. For now, it's enough to consider that almost every object in the physical world

---

[4] This is also what Dan Roam of *The Back of the Napkin* fame has been advocating for a few years now. At the 11th ASIS&T IA Summit in Phoenix, Arizona, Dan showed the audience how turning the apparently complicated into the complex but logical actually could make the recent U.S. health care reform bill something to discuss on its merits rather than on ideological grounds (Roam 2009).

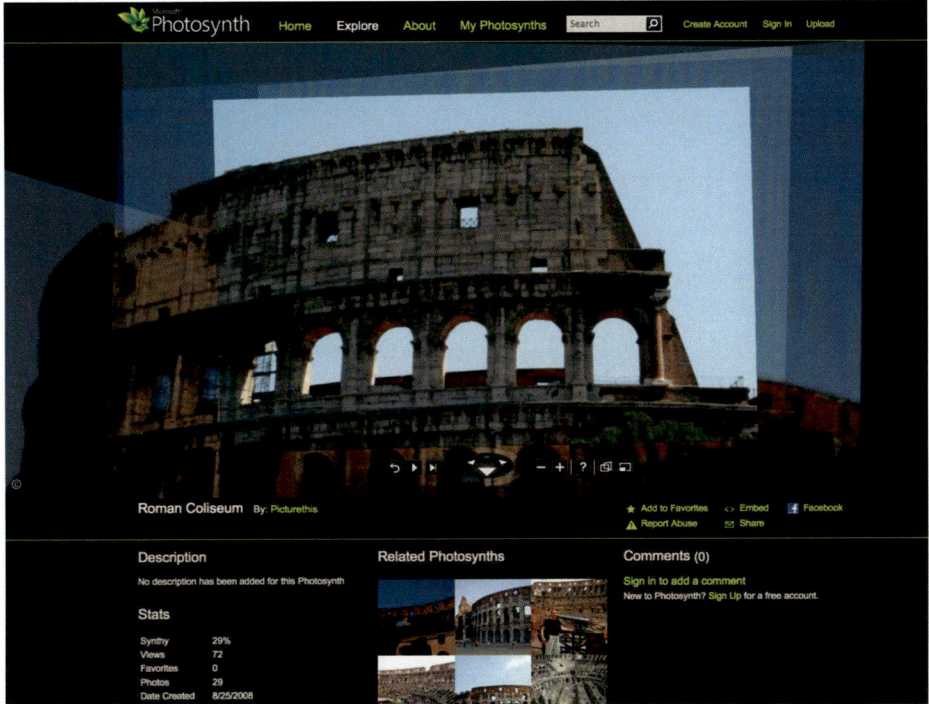

FIGURE 6.8

The Coliseum, Rome, in a Photosynth 3D reconstruction from user-generated photos. *Source: Photosynth.net*

projects into the digital world what Kuniavsky (2010) calls an *information shadow:* performing loops of simple operations such as monitoring, filtering, and reusing on those shadows can provide some of the foundations and understanding upon which resilient environments can be built. We do not need explicit tracking systems or techniques to do that: our interactions with people, places, objects, and information produce a large amount of data even if these have no unique machine-readable id. And this is not limited to the latest, coolest gadgets you can buy.

Stop for a moment and think of the Coliseum. Think of its *shadow* on the Internet: geotagged Flickr photos taken by people you know or don't know with cameras and smartphones, pages, and links mentioning them on either Wikipedia or a thousand blogs and online newspapers. Tweets, comments, diggs, videos, podcasts. Or think of your house. Of course the Coliseum will cast a much larger shadow, at least unless you are Barbra Streisand,[5] but chances are you tweeted about your whereabouts, made it into a spot in Gowalla! or Foursquare or any other geo-social mobile game, or blogged about some event that happened there. If you didn't, someone else did, if only because they live next door and send pictures to a relative living in South Africa. Triangulating

---

[5] If you don't know what we are talking about, just check Wikipedia for the Streisand effect.

such data allows for a crude but rather effective way to track the digital life of real-world objects: we all have a favorite Web site that does precisely that, from the small blog that collects all available information on local restaurants to Microsoft Photosynth and its 3D reconstructions of the world by means of user-generated photos on Flickr (Figure 6.8). But the real deal is when this information moves out of the Web, when we start feeding the digital into the physical, when we augment reality with pioneer consumer applications that sustain user-centered information consumption and information production. These are pipes simultaneously feeding and feasting on the vast pool of information shadows already available. They make the wheel turn.

> As the information shadows become thicker, more substantial, the need for explicit metadata diminishes. Our cameras, our microphones, are becoming the eyes and ears of the Web, our motion sensors, proximity sensors its proprioception, GPS its sense of location. Indeed, the baby is growing up. We are meeting the Internet, and it is us.
>
> (O'Reilly & Battelle 2009).

The more digital and physical overlap and converge, the more we contemplate "a nature from top to bottom written," as Foucault wrote in *The Order of Things*. Our job is to find a way to make this writing accessible, meaningful, and resilient.

LESSONS LEARNED

Know

- Resilience makes an information space able to adapt itself to the changing needs of its users in different contexts of use, different places, and different times.
- Resilience makes an information space capable of supporting multiple information seeking strategies, either active or passive, directed or indirected, conscious or latent.
- Places are texts. Places are palimpsests where people write and rewrite their interactions with the environment, with other people, or with objects.
- Most objects leave traces and project shadows in information space.

Do

- Integrate bottom-up, user-created patterns with top-down, built-in structures to improve the resilience of an information space.
- Make these two levels communicate: allow fast but consolidated user-created patterns to seep down to the foundations and allow slow structures to be moved, changed, and flexed when needed.

■ Collect, filter, and reuse the traces and shadows objects and people leave in information space to allow users to satisfy their natural propensity for harvesting pieces of information passively and to elicit latent needs.

CASE STUDIES

The Resilient Museum

Save for highly specialized cases, collections always verge on the incongruous. A space traveller unaware of our concept of art would wonder why the Louvre contains trifles in common use such as vases, plates, or salt cellars, icons of a goddess such as the Venus de Milo, representations of landscapes, portraits of normal people, grave goods and mummies, portrayals of monstrous creatures, objects of worship, images of human beings suffering torture, accounts of battles, nudes calculated to arouse sexual attraction and even archeological finds.

(Eco 2009, p. 169).

Museums used to work like that. It was a rather common experience for us who grew up in the 1970s: visiting a museum was like trespassing into a silent, sacred space, where the clicking of heels sounded like thunder and where we stared in awe and wonder at what seemed to us simply very well-organized disorder. The Roman rooms on the right, the Egyptian rooms on the left, the Middle ages somewhere on the second floor, and the souvenir desk somewhere close to the exit (Figure 6.9). The only persons to have an idea of what was going on were the teachers, who planned their visit in advance with the aid of books. And to be totally honest, that was part of their duties.

For almost all of us, and for most of the time, it was like having the Chinese list read to us while following brother William through the maze of the library (see Chapter 4, p. 79): bewildering. We were like customers in our Crowded Winery; as we had no precise idea of what we wanted (that famed known-item seeking behavior), we could make no sense of the wild array of rooms, items, and stories in front of our eyes.

At rare times, though, something magical happened. You could visit a museum that hosted a collection of something you were into, say, dinosaurs, or, well, dinosaurs.[6] Then everything changed: you were the one running around, explaining this and that to classmates, teachers, and guides alike. How was that?

---

[6] As wax museums did not really count and superheroes were not really top of the list in the mind of many a curator.

FIGURE 6.9
How museums used to be.

Through the years, and through the toil of countless armies of increasingly bored and restless classes, museums have become more comfortable and welcoming places. They even have coffee bars and recreational areas.[7] Current museums are less and less the temple and more and more the forum, meeting centers where comparisons, research, and debate take place, and where visitors can interact with the exhibitions, enter their stories, and, to some extent, modify them. Museums have become integrated, playful, discover-friendly experiences; in Renzo Piano's NEMO Science Center (Figure 6.10) on the waterfront of Amsterdam, The Netherlands, visitors are constantly engaged in activities that invite them to "smell, hear, feel and see how the world works." Learning about the origin of life, how electricity works, or what is DNA is part of a dynamic experience that would send some of our old-school teachers from the 1970s screaming.[8]

Of course, NEMO is not the only museum of this kind: in Oslo, Norway, you can visit the Norsk Teknisk Museum and build your own rocket or experience climate change consequences first hand. Your own city might have some similar installation. It is important to note that this is not just some playful

---

[7] We don't know about the rest of the world, but some law was certainly in place in 1970s Italy to forbid that. No bars allowed. No playgrounds. And if we have to judge by the way it was obeyed, it is safe to assume breachers were sentenced custody for long years or to working the mines.

[8] NEMO can be found online at http://e-nemo.nl. Check out this YouTube video for an idea of how children can experience science there: http://www.youtube.com/watch?v=huh7U82Hezk.

FIGURE 6.10
How museums are.
The NEMO Science
Center in Amsterdam,
The Netherlands.
Photo: J. Nieuwland.

experience designed exclusively for children, as the interactive (and graphical) nature of the permanent and temporary exhibitions at the Apartheid Museum in Johannesburg, South Africa, clearly exemplifies (Figure 6.11).

Now there's an interesting observation: museums host artifacts that, because of time, cultural reasons, or artistic considerations, present a complex nature, offer multiple meanings, and are generally susceptible to be interpreted differently by different audiences (Sbicca 2009). They have many facets. As such, the most effective museums are those that allow for full, active participation of the visitors and that stimulate a playful, emotionally grounded experience: emphasis shifts from conservation and exposition to interaction.

Neil, Philip, and Wendi Kotler define six different types of museum experiences: recreative, socializing, learning, aesthetic, celebratory, issue oriented, and enhancing (Kotler et al. 2008, p. 303). Visitors have different goals: some want simply to be able to finally see the one masterpiece they have long seen in books, some want to be surprised and amazed, some are just curious, and some try to connect the dots that link different artists, paintings, or cultures. A resilient museum should be able to speak to all of them in different ways (Figure 6.12).

However, the arrangement of exhibits and artwork in the museum is inevitably bounded by physical constraints: each piece can be in just one place, as much as the bottles in the Crowded Winery. Cloning items, which is what a large shop might decide to do in the slightly less technological version of placing the same product in different shelves, is not an option for a museum: you cannot really have more than a thousand Warhols. Sorry, we meant one *Mona Lisa*.

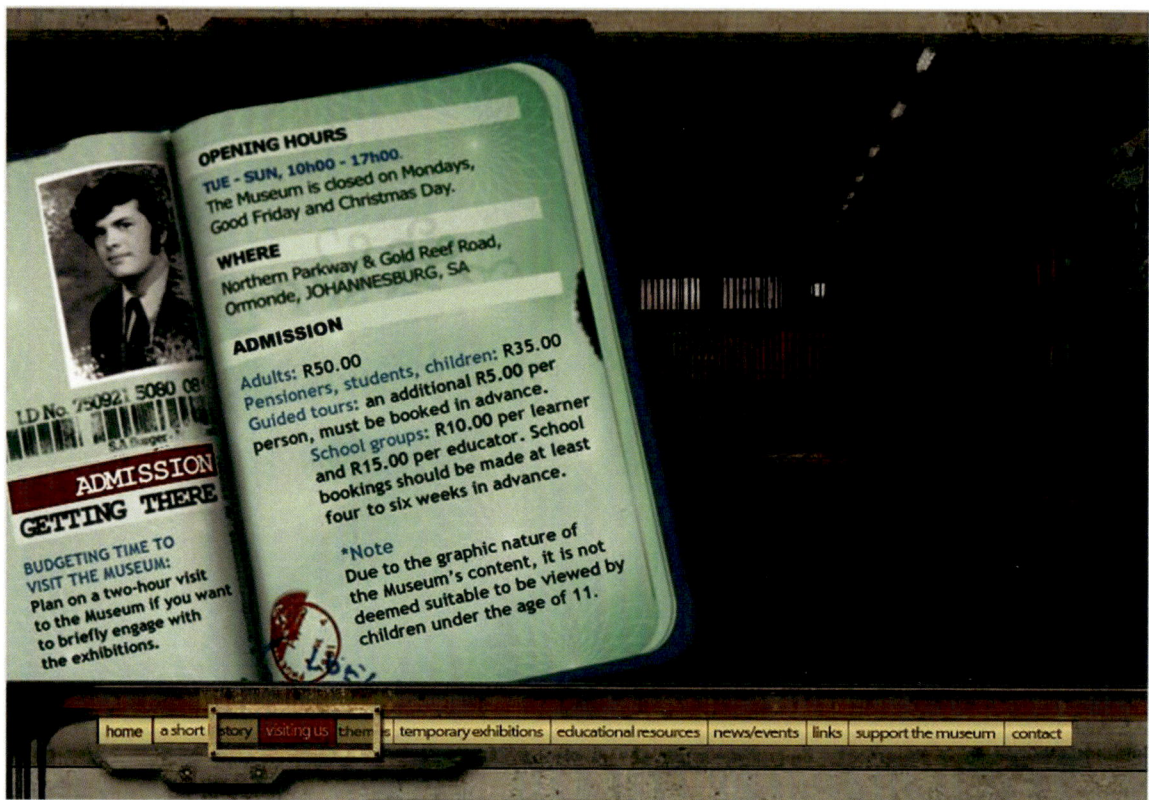

FIGURE 6.11

The Apartheid Museum in Johannesburg Web site.

How to design for those different visitors then? How to make the museum resilient?

What we have been saying in this chapter about tracking interactions and hypertextual links in physical environments offers a solution: even if the items and their locations are unique, the fact remains that they can be semantically connected and physically reached through multiple, different paths. This is the logic that many libraries implement: even if books have unique locations, they can be searched using several different parameters. After all, faceted classification was conceived for the physical environment of a brick-and-mortar library. What if we stretch this idea again across channels and environments?

In their work for the EU SHAPE (Situating Hybrid Assemblies in Public Environments) project, devoted to "the design of hybrid room-sized installations that merge together physical and digital elements in meaningful ways," Luigina Ciolfi and Liam Bannon (2002) of the University of Limerick, Ireland, explored how visitors experience the museum according to the

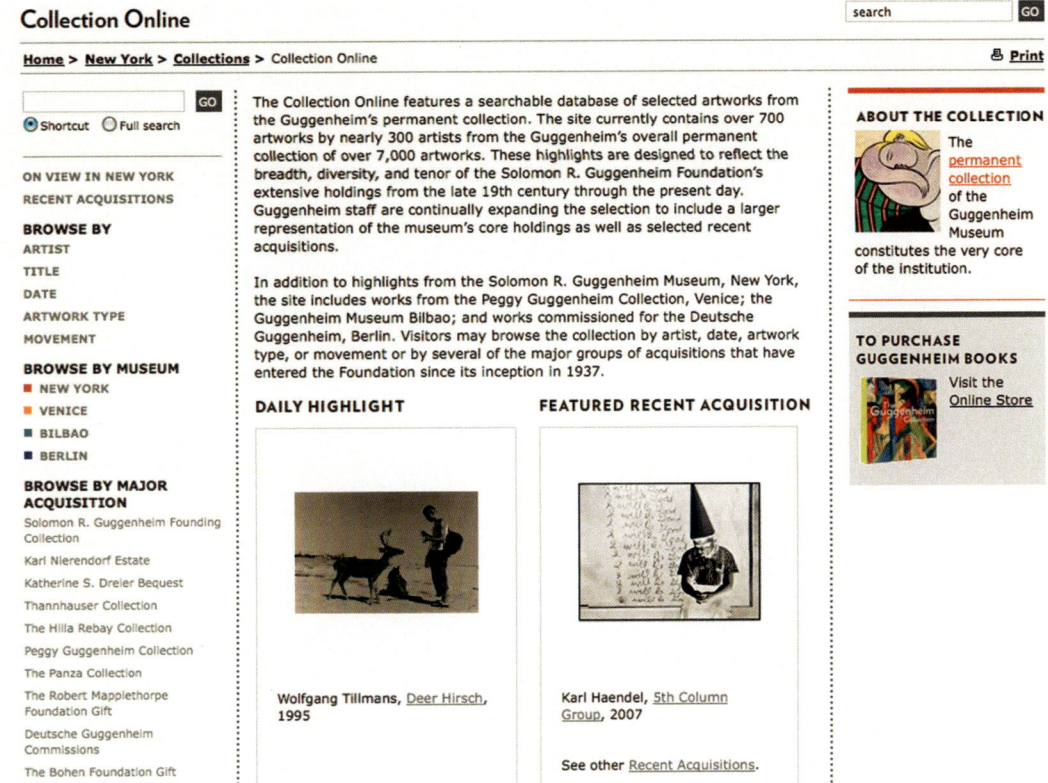

FIGURE 6.12
Home page for the Guggenheim Museum in New York, showing a mixture of facet-like static navigation (on the left) and dynamic navigation (in the body).

perspective of "localised experience," with the goal of informing the design of a novel interactive exhibition space. Ciolfi and Bannon believe that "gaining a thorough understanding of the way visitors move through the exhibitions and interact around the objects on display is a crucial element in designing effective museum installations": visitors were video recorded, interviewed, and observed. The researchers especially traced how they moved around and how they interacted with artifacts and with other people. This was not just analysis for the sake of research, though: after all, SHAPE exhibitions

> should allow visitors to leave a trace of their visit and their interaction with the objects, something which might in turn shape and influence the experience of other, later, visitors to the Museum.

What that means is that findings from these early experiments were used to improve the design of a final exhibition, where participants were actively involved in the exploration of less-known objects from the collection, and were able to leave traces that potentially influenced other people's experience of the exhibit.

At its simplest level, a resilient museum could for example integrate three classification layers, three different mixable strategies for shaping visitors' interactions, working at different speeds according to **pace layering** theory: a stable top layer, relatively slow; a second, more fluid layer, relatively fast; and a third layer acting as a service and dialog layer between these two (Table 6.4). These layers are then usually embedded in different, channel-specific artifacts.

In such a scenario, the top-down level might employ a faceted classification scheme or a faceted thesaurus such as the *Art & Architecture Thesaurus* used by the Getty Trust. A faceted approach is easily the best solution to help visitors build their own paths by assembling different facets (such as all *French paintings* from the *Renaissance*) according to their personal needs such as Lego bricks and to allow the staff to build temporary thematic paths or suggest related items of interest/paths.

The bottom-up layer consists of the interactions and paths drawn by visitors using the museum's information system (built on the top-down layer), their personal tastes, culture, and needs or any external connected source (a book or a social tagging system). This layer is fast paced, rapidly changing, and largely unpredictable.

Finally, the middle layer, the up-and-down layer, is a service layer mediating between top and down. Its role is to balance the way information from the bottom can be progressively promoted to the top. Recurring and frequent bottom-up patterns are used to fertilize and optimize the top-down

Table 6.4 Simple Scheme of Integrated Classification Layers in a Museum

Layers	Classification System	Enabling Tools and Technologies
Top down	Faceted classification system Thematic paths	Analogical and/or digital signage employing alphanumeric and chromatic codes. Searchable digital catalogs
Bottom up	Social tagging and navigation: personal/social paths created by people using facets, tags, and their own competencies	PDAs, smartphones, GPS, RFIds or similar systems. Augmented reality systems
Up and down (service layer)	General monitoring and controlled absorption into the top-down layer of the paths and stories created by the visitors	Monitoring and measurement systems logging and reporting repeating paths and behaviors

level, improving the general information architecture, introducing new facets or terms, and ruling out unnecessary or underused ones. This idea of a middle mediating layer has been explored thoroughly by the BBC with their *metadata threshold* discussed in the next case study.

> In some way, then, the cumulating frictions of the cinema and the architecture of film theaters have come to reinvent . . . some of the imaginative process that, in 1947, André Malraux called the musée imaginaire: a boundless notion of imaginative production that, in English translation, becomes "a museum without walls." . . . In this vein, a cultural landscape, broadly conceived, can be regarded in many ways as a trace of the memories, the attention, and the imagination of those inhabitant-passengers who have traversed it at different times. It is an intertextual terrain of passage carrying its own representation in the threads of its fabric, weaving it on intersecting screens. A palpable imprint is left in this moving landscape; in its folds, gaps, and layers, the geography of cinema and the museum holds remnants of what has been projected onto it at every *transitio*, including the emotions. Imaged in this way, such a landscape is an archaeology of the present.

(Bruno 2007, pp. 33, 39).

Pace layering - Pace layering is a concept originally introduced by Stewart Brand, an American designer and writer, in his 1994 book *How Buildings Learn* to explain how different rates of change affect buildings. Brand maintains that buildings can be seen as organized into different layers that change at different speeds; expanding on architect Frank Duffy's ideas, he built a *six S's* model composed of site, structure, skin, services, space plan, and stuff. While site, the geographical location, is "eternal," all the other layers are in constant flux, the structure very slowly, the skin faster, the services even faster, the space plan faster again, and the stuff, interiors and furniture, fastest. Through various revisions of the initial idea, Brand came to formalize pace layering as a central element in the development of generic complex systems, with complexity being the result of dialogue between the fast layers, which bring in innovation, and the slow layers, which provide continuity and stability.

The BBC and the Metadata Threshold

In her article "Changing Approaches to Metadata at bbc.co.uk: From Chaos to Control and Then Letting Go Again," British information architect Karen Loasby (2006) describes the evolution of metadata management in the context of the British Broadcasting Corporation (BBC, the public TV broadcaster in the United Kingdom) Web site since 2002, providing emblematic evidence to the arguments explored in this chapter: mixing together a top-down and bottom-up approach, monitoring human–information interactions for further reuse allows the designer to vastly improve the resilience of an information space.

For many years the Web Division staff at the BBC believed that metadata could be the solution for providing visitors to the Web site with some useful helping guide, but they somewhat failed to find the right way to implement this.

As of 2002, the use of metadata in the Web site was still limited to adding keywords to pages for improving indexing by both the internal search engine and Google.

In 2004 there was a change in strategy, and focus shifted: it was thank you Google, it's been nice, but we need to move up the ladder. Metadata started to be perceived primarily as an instrument for improving the distribution and aggregation of content, and in this perspective its correct and unambiguous use became essential to avoid dubious classifications or incongruous correlations of the video materials hosted on the Web site.

As we know very well, dubious, ambiguous, and incongruous are second names to any language and labeling system: so the staff decided they needed to add some degree of order to this structure before the situation got out of hand and introduced controlled vocabularies. They soon found out that the supposedly simple task of maintaining and updating the controlled vocabularies keeping in check the vast range of BBC productions was in fact a dire, hard job. As a consequence, their application was limited to just those sections of the site that were centrally managed via the internal content management system, preventing or severely crippling, then, any meaningful aggregation of content across the entire site. Which kind of defeated the purpose, but nobody says IAs have it easy every time.

Then, two years later, in 2006, there was a second breakthrough: the idea that content classification through metadata might be vital and strategic for the Web site filtered up to management. A decision was made to extend the use of controlled vocabularies and, consequently, to solve any managing and updating issue they could pose. As a system relying on centralized governance was ruled out quickly because of the high costs associated in terms of both time and staff, the idea of combining a top-down taxonomy (by means of the controlled vocabularies) with a bottom-up classification created directly by users via tagging[9] started to acquire momentum. Finally, the project was green-lighted and met with success.

Now, user-generated taxonomies, folksonomies as they have been brilliantly and aptly named by Thomas Vander Wal, naturally carry with them a certain inevitable disorder[10]: therefore the tags posted by users cannot be incorporated into the controlled vocabulary without an appropriate "treatment" or without being sanitized. To solve this issue, the BBC Web Division staff conceived a middle layer called the *metadata threshold*, whose only purpose is to mediate the dialog between bottom-up terms and the top-down, more formal structures (Figure 6.13): very pragmatically, all tags exceeding a preset threshold frequency

---

[9] See BBC Backstage and Cambiassi (2006).
[10] For more on folksonomies, see Quintarelli (2005) and our own FaceTag project (2006–2008).

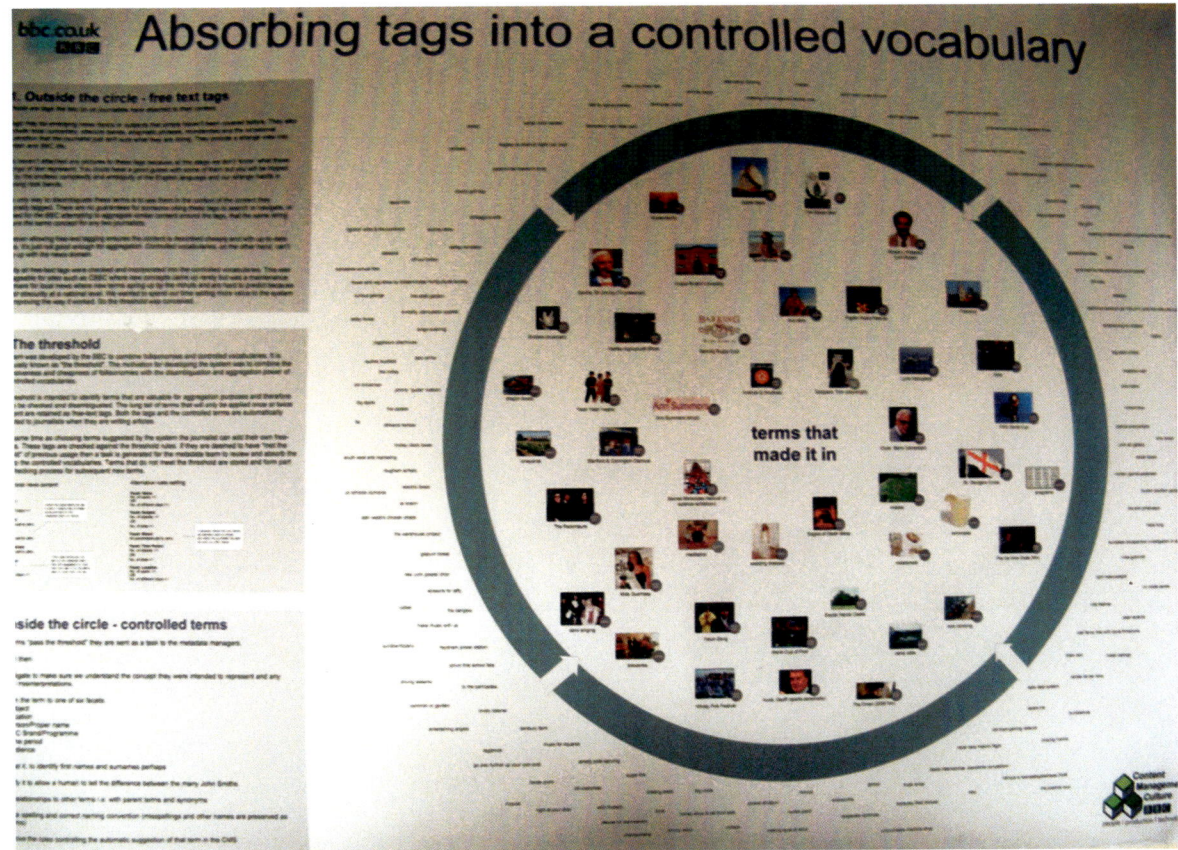

FIGURE 6.13

The metadata threshold: Combining folksonomies and controlled vocabularies at the BBC. Poster presented at the 2nd European Information Architecture Summit (Berlin 2006). Photo: K. Loasby.

are absorbed into the controlled vocabulary and contribute to modifying the initial state of the system, producing change.

By exploiting the primary benefit of folksonomies, having the users themselves contribute to the classification, this solution helped reduce the amount of work weighing in on the staff and contributed a low-cost pragmatic answer to the management of a complex information system. By integrating a top-down, centralized and rigid taxonomy with a bottom-up folksonomy run directly by the users, and therefore much more open, it brought the best of both worlds to the final product, as each system balanced and corrected the limitations of the other.

Finally, by involving users in the site management process it created participation, opened up an important feedback channel for the BBC, reinforced its brand, and probably helped understand and perhaps anticipate the needs and tastes of the audience itself.

RESOURCES

Articles

Agarwal, S. (2009). Building Rome in a Day. http://grail.cs.washington.edu/rome/.

Bates, M. (2002). Toward an Integrated Model of Information Seeking and Searching. *New Review of Information Behaviour Research, 3,* 1–15. (Available at http://www.gseis.ucla.edu/faculty/bates/articles/info_SeekSearch-i-030329.html).

Campbell, D. G., and Fast, K. (2006). From Pace Layering to Resilience Theory: The Complex Implications of Tagging for Information Architecture. In *Proceedings of the 7th Information Architecture Summit.* Vancouver, March 23–27. ASIS&T. http://www.iasummit.org/2006/files/164_Presentation_Desc.pdf.

Ciolfi, L. (2004). Understanding Spaces as Places: Extending Interaction Design Paradigms. *Cognition Technology & Work, 6*(1), 37–40.

Ciolfi, L., & Bannon, L. (2002). Designing Interactive Museum Exhibits: Enhancing Visitor Curiosity Through Augmented Artefacts. In Bagnara, et al., *Proceedings of ECCE11, European Conference on Cognitive Ergonomics.* Catania, September 2002. http://richie.idc.ul.ie/~luigina/PapersPDFs/ECCE.pdf.

Loasby, K. (2006). Changing Approaches to Metadata at bbc.co.uk: From Chaos to Control and Then Letting Go Again. *Bulletin of the American Society for Information Science and Technology, 33*(1), October/November. http://www.asis.org/Bulletin/Oct-06/loasby.html.

Marchionini, G. (2004). From Information Retrieval to Information Interaction. In S. McDonald and J. Tait (Eds)., *Advances in Information Retrieval: 26th European Conference on IR Research, ECIR 2004, Sunderland, UK, April 5–7, 2004, Proceedings* (pp. 1–11), Springer. http://ils.unc.edu/~march/ECIR.pdf.

Merholtz, P. (2009). Desire Lines. Adaptive Path Blog, October 27, http://www.adaptivepath.com/blog/2009/10/27/desire-lines-the-metaphor-that-keeps-on-giving.

Quintarelli, E. (2005). Folksonomies: Power to the People. In *Proceedings of ISKO Italy Meeting.* Milan, June 24. http://www.iskoi.org/doc/folksonomies.htm.

Quintarelli, E., Resmini, A., & Rosati, L. (2008). The FaceTag Engine: A Semantic Collaborative Tagging Tool. In Zambelli et al. (Eds.), *Browsing Architecture: Metadata and Beyond* (pp. 204–217). Fraunhofer IRB.

Spencer, D. (2006a). Four Modes of Seeking Information and How to Design for Them. Boxes and Arrows, March 14. http://www.boxesandarrows.com/view/four_modes_of_seeking_information_and_how_to_design_for_them.

Williamson, K. (2005). Ecological Theory of Human Information Behavior. In Fischer et al. (Eds.), *Theories of Information Behavior.* (pp. 128–131). Information Today.

Books

Bachelard, G. (1969). *The Poetics of Space.* Beacon Press.

Brand, S. (1994). *How Buildings Learn.* Penguin Books.

Rosenfeld, L., & Morville, P. (2006). *Information Architecture for the World Wide Web.* 3rd ed. O'Reilly.

Snavely, K. N. (2008). *Scene Reconstruction and Visualization from Internet Photo Collections.* PhD Dissertation. University of Washington. http://grail.cs.washington.edu/theses/SnavelyPhd.pdf.

Sterling, B. (2005). *Shaping Things.* MIT Press.

Videos

Barlow, E. (2010). How Complexity Leads to Simplicity. *TEDGlobal*. http://www.ted
.com/talks/eric_berlow_how_complexity_leads_to_simplicity.html.

Klima X. *Norsk Teknisk Museum*. http://www.tekniskmuseum.no/film-videoklipp/fra-utstillinger/
klima-x

What Is Science Center NEMO? http://www.youtube.com/watch?v=eWpA9Dihc30.

CHAPTER 7

Reduction

FIGURE 7.1

LUCA'S BIG ADVENTURE WITH A DO-IT-YOURSELF SCALE

I'm in the neighborhood supermarket for my weekly shopping expedition. My cart is loaded with fruits and vegetables, and I approach one of the many do-it-yourself electronic scales to weigh the different parcels and get their costs in order before lining up. As usual, when I get to the scale I already forgot the codes associated with some of the things I'm buying so I have to go back to the stands and look up their numbers again. The system works by associating any fresh fruits or vegetables with a specific number: you then have to choose that number or punch it in on the scale.

This association, however, is not stable, as it changes with the normal turn-over of fresh produce. In the winter, a certain number might be oranges and in the summer peaches or watermelons. The scale has a large touch-screen with buttons: even when each button presents an icon beside the number, their smallish size and the large number of items make this a dire task. Plus, steering an overloaded cart around is far from easy: I'm tired like everybody else and want to be home.

I resolve to leave the cart by the scales and walk back to read the numbers, but it seems that I'm not the only one to have some short-term memory issues: there is a constant flow of people moving back and forth. I start to notice that some of them have figured out workarounds, or complete, full-blown strategies. For example, a couple with one rather turbulent child has developed some sort of memory game: remembering the numbers and pressing the right button on the display has become a playful experience, and they even get the child not to wander around in the process. An older couple who can obviously profit from a higher number of children has created a short assembly line: one fetches the fruit or vegetables, one reads the number, and one manages the scale. Two friends are somewhat less creative but not less effective: she chooses her grocery carefully and hands some to him. While she moves to the next aisle he moves to the scale and shouts back something like "melon!" to which she replies "34" in a very loud voice. They really have some troubles judging distances, as they seem to be ordering troops over a battlefield, and they repeat the procedure until they are finished. Everyone is glad they apparently do not eat that much green stuff.

All of this moving around, playing, calling, and shouting turn the supermarket into a slightly toned-down clone of a traditional street market, with some air conditioning thrown in. As so often happens, instead of making things simpler, technology, or those who design it, has just made my life and that of my fellow shoppers trickier. It transformed a simple task (getting a price tag by weighing a few vegetables) into an adventure. If all you are looking for is some opportunities to entertain your children, you might find something useful here, I agree. However, if you are shopping late in the evening after a day's work and can't wait to be home, then you are not going to appreciate. Anything that has you walk back and forth 10 times, making you ponder whether you are developing some serious cognitive problems all along can't be right.

THE ROOT OF THE PROBLEM

Reduction - The capability of an information space to minimize the cognitive load and frustration associated with choosing from an ever-growing set of information sources, services, and goods. It is also a set of strategies to address such an issue and has nothing to do with getting rid of choices: reduction is an organizational and presentational guideline.

There is no doubt that the design of these do-it-yourself scales leads to cognitive load and to psychological stress. Nevertheless, even if this might apparently seem just a memory-related issue, its real cause lies elsewhere. The cognitive effort we exert in trying to remember the product number is mostly due to the difficulty of finding the matching icon or button on the scale's display. And if you think this is again just the result of having too many products from which to choose you are being misled. The cognitive load has not as much to do with the large numbers of options as with how these choices are organized and presented. **Reduction** is a way to address this design issue.

LONG TAILS AND CHOICE OVERLOAD

Product and service production is moving toward what Chris Anderson has dubbed the **long tail** model (Figure 7.2): strong differentiation and personalization, attention to an increasing number of niche markets, and the willingness to sell fewer items to many instead of many items to a few (Anderson 2006). Among other things, this implies larger catalogs and more information available.

As if this was not enough, the convergence between digital and physical and the whole Internet of Things phenomenon are generating an overwhelming quantity of data: objects produce positioning info, status message, logs, notifications, and statistics. Most of these end up on the Web and create what Mike Kuniavsky calls the *information shadow* for a specific object. We introduced this concept in the previous chapter when discussing resilience.

Then choice becomes a difficult enterprise. Choosing among an ever-growing number of possible brands, models, or sizes for any single item we buy or use can be a most stressful experience: this is what Richard Saul Wurman (2000) and Barry Schwartz (2005), respectively, call **information anxiety** and **paradox of choice**. Both definitions nail down a distinctive aspect of the issue at hand: we can certainly become anxious because of the excessive number of choices from which to discriminate, but, and here lies the paradox, such abundance is a richness and a slightly intoxicating habit we would not be able to renounce easily. Once you have much, it is psychologically difficult to go back to have little.

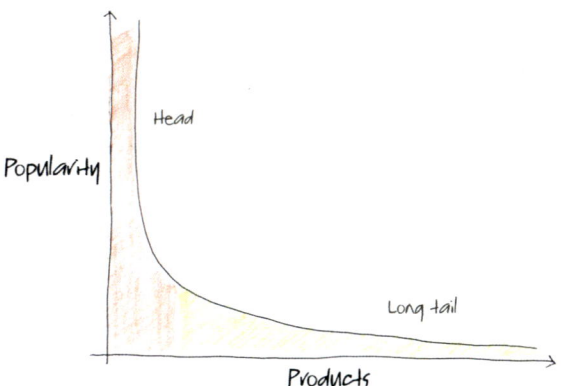

Choice and the long tail

FIGURE 7.2

The long tail model: the *y* axis represents the sales for a given product (popularity); the *x* axis represents the different products being sold. The head is where best-selling items are; the long tail is where the niche markets are.

Behavioral Economics - How people decide is one of those topics that get a lot of attention from many different disciplines. Economics, psychology, philosophy, mathematics, and statistics have all formulated several theories and models to understand how decision making works. Behavioral economics is one of these disciplines, and it brings in social, cognitive, and psychological concerns in trying to understand the economic decisions of both individuals and organizations.

So, how could we conciliate these opposite needs, that of a long tail economy and personalized seeking strategies on the one hand and that of information overload and too many choices on the other? Apparently, we have a conflict: users need more information, but more information can easily become less information (Figure 7.3).

FIGURE 7.3
Choosing your afternoon tea could prove difficult.

One thing we can say: the problem is not complexity itself or the sheer number of different exotic vegetables with unpronounceable names the supermarket can put on their shelves and onto the scales' displays. We said it before, but it's worth repeating: complexity is a form of richness that needs to be exploited; the abundance of products is a fact of life. The problem lies in the lack of any effective information organization strategy governing the process.

Italo Calvino on lightness and multiplicity

Italian writer Italo Calvino wrote about **lightness** and **multiplicity** in his famous American lectures, *Six Memos for the Next Millennium*:

> my working method has more often than not involved the subtraction of weight. I have tried to remove weight sometimes from people, sometimes from heavenly bodies, sometimes from cities.... I have come to consider lightness a value rather than a defect.
>
> (Calvino 1993, p. 3).

But Calvino does not advice for easy escapism. Reality is complex, and representing this complexity is the challenge:

> to represent the world as a knot, a tangled skein of yarn; to represent it without in the least diminishing the inextricable complexity or, to put it better, the simultaneous presence of the most disparate elements that converge to determine every event.
>
> (Calvino 1993, p. 106).

Simplicity and complexity complement each other

What Calvino is saying is that there is no dichotomy here, no challenge that can be solved by cutting out, leaving out, or discarding. Being light or simple *and* multiple or complex at the same time are the final goals. These different qualities are not mutually exclusive: in their **unbalance**, design thrives.

The solution for better user experience in cross-channel information spaces does not consist in the simple eradication of options–at its worst coalescing into some utopian going back to mythical, unspoiled, simpler origins where only the "right" choices were available–but in working at ways to improve human–information interaction all along the process.

We could say it is not an issue of quantity but rather an issue of quality, as it relates directly to how information is organized, visualized, and finally made accessible: considerations on place, time, and context while delivering information are paramount. This is how we intend it: reduction does not stand for fewer choices and does not stand for trying to wish complexity away like it was black clouds rolling in a summer sky, but for a design mindset that strategically helps reduce an excessive, useless simultaneous number of choices into a manageable, meaningful flow across channels.

WHEN MORE IS LESS

Filtering out extraneous information is one of the basic functions of consciousness. If everything available to our senses demanded our attention at all times, we wouldn't be able to get through the day.

(Schwartz 2005, p. 23).

Schwartz correlates the stress brought on by an excess of choices with **locus of attention**, our current focus, whose essential property is its singularity: we have one and only one locus of attention and there is no way to activate a second one. That is to say: we cannot pay attention—voluntarily or involuntarily—to more than one item a time (Raskin 2000, p. 24; Figure 7.4).

Locus of attention

Examining the results of some empirical studies, Schwartz explores other cognitive mechanisms that seem to explain why choosing is a stressful activity, some of them closely resembling the *principle of least effort* we introduced when discussing *resilience*. This basic principle of getting the most out of the littlest of toils can be said to influence large parts of human behavior, from

FIGURE 7.4
Too many options at hand frustrate even aliens. Pixar's *Lifted* (2006).

language to information seeking: coupled with the anxiety that comes from being unsure of what the results of our choices will be (wouldn't it have been better if I had chosen the other phone or the other car?), this seems to be the root of much of the stress we suffer while choosing.

> Apparently, people are shopping more now but enjoying it less....
> A recent series of studies, titled "When Choice Is Demotivating" [see Iyengar and Lepper 2000], provide the evidence. One study was set in a gourmet food store in an upscale community where, on weekends, the owners commonly set up sample tables of new items. When researchers set up a display featuring a line of exotic, high-quality jams, customers who came by could taste samples, and they were given a coupon for a dollar off if they bought a jam. In one condition of the study, 6 varieties of the jam were available for tasting. In another, 24 varieties were available. In either case, the entire set of 24 varieties was available for purchase. The large array of jams attracted more people to the table than the small array, though in both cases people tasted about the same number of jams on average. When it came to buying, however, a huge difference became evident. Thirty percent of the people exposed to the small array of jams actually bought a jam; only 3 percent of those exposed to the large array of jams did so. . . . The authors of the study speculated about several explanations for these results. A large array of options may discourage consumers because it forces an increase in the effort that goes into making a decision. So consumers decide to not decide, and don't buy the product. Or if they do, the effort that the decision requires detracts from the enjoyment derived from the results. . . . Having too many choices produces psychological distress, especially when combined with regret, concern about status, adaptation, social comparison, and perhaps most important, the desire to have the best of everything—to maximize.
> (Schwartz 2005, pp. 19–20, 221).

Diminishing marginal utility

This problematic conundrum finds its possible explanation in the law of **diminishing marginal utility** (Schwartz 2005, pp. 67–73).

In economics, the marginal utility of a good or service is the additional utility gained (or lost) from an increase (or decrease) in the consumption of that good or service. We could say that marginal utility is the benefit provided by the acquisition of further amounts of a given good or service: the law of decreasing marginal utility states that a good or service's marginal utility decreases as their quantity (in our possession) increases while the consumption of other goods or services does not change.

In other words, if we buy something, the first batch is going to provide us with more value than, say, the third or the fourth one. Everyone but an economist would expect that the relationship between the utility of some good or service and its utility to us is proportional to the quantities of that good or service we possess, but this isn't the case: the function of the law (Figure 7.5) is a curve whose slope diminishes as the quantity increases.

An example might help us better understand why it is so. Imagine that you are sitting at an elegant restaurant having dinner. You find a magnificent bottle of Amarone della Valpollicella on the table, perfectly arranged in its decanter and served in large Burgundy glass balloons (Figure 7.6). The first sip or glass will give you wonderful sensations and will give you a benefit you score 10 out of 10. The second glass will certainly give you great sensations again, but it will score lower, say, 8 out of 10. If you are like Andrea and are no wine lover or connoisseur, take Luca's word on this: drinking such a wine for the first time is incomparable. As your evening moves on, drinking more of the wine will get fewer and fewer points in your personal score. Plus, you will also have to stop if you do not want to get drunk.

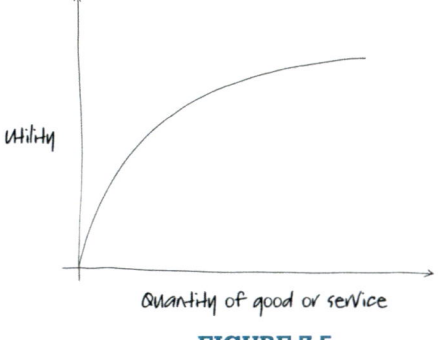

FIGURE 7.5
Diminishing marginal utility.

Not only that, but once we go past being fully gratified, any increase in acquisition or consumption (wine, in our example, but it could be any other thing, including fashionable clothes or electronic gadgets) will probably result in **disutility**, having a diminishing effect on the individual level of satisfaction. At the exact tipping point, when we are sated, marginal utility is null and we are indifferent to having or not having another glass of wine. Here utility is maximized.

From utility to disutility

Let's consider information like it was any other good: what happens if we apply this idea of marginal utility to the process of choosing? Well, the law of decreasing marginal utility demonstrates that the increase in the amount of information, and consequently in the number of available choices and in the quality of our judgment, increases our satisfaction nonproportionally, and only up to a certain threshold that coincides with the point of satiety. Once the

FIGURE 7.6
Drinking Amarone as an example of diminishing marginal utility applied to wine consuming.
Photo: E. Ziliani. *Source: Vigneti Villabella.*

Hick's law

information available or the number of options exceeds the threshold, their marginal utility begins to decrease.

If you think you had your share of principles, curves, and theories, we are afraid you will have to endure a little more of this: we have to introduce yet another important law that concerns itself with choice-related issues: **Hick's law**. The good part is that Hick's law and the law of diminishing marginal utility perform similarly and produce comparable results.

HICK'S LAW

Hick's law, also known as the *Hick–Hyman law,* after the names of the psychologists William E. Hick and Ray Hyman who formulated it in the 1950s, provides a mathematical model to understand the paradox of choice and suggests some countermeasures to reduce its impact. Get ready and read it in one breath: given *n* equally probable choices, the average reaction time required to choose among them is approximately proportional to the logarithm to base 2 of the number of choices, plus 1. If we want to formalize it, we can write that as

$$\text{time} = a + b \log_2 (n+1),$$

where *a* and *b* depend on context conditions, such as presentation and the user's degree of familiarity with the subject. For example, if choices are presented poorly, both *a* and *b* increase, while familiarity only decreases *b*.[1] Pause. We don't know about you, but we must admit it took us some time to actually wrap our minds around it. Let's try to make it more human-friendly by means of an example, or more than one.

In their book *The Art of UNIX Usability*, Eric S. Raymond and Rob W. Landley (2004) paraphrase and explain Hick's law as

> the time $M(n)$ required to make a choice from a menu of n items rises with the log to the base two of n. The key fact here is that the rise of $M(n)$ is sublinear. Thus, the Rule of Large Menus: one large menu is more time-efficient than several small submenus supporting the same choices, even if we ignore the time overhead of moving among submenus.

Now consider *menus* as a catch-all word for a generic way to show choices. What's worth noticing is that the correlation between reaction time and available choices is expressed logarithmically, that is, it is nonlinear (Figure 7.7). We could indeed expect that the relationship between time and the number of

---

[1] For an overview, see Raskin (2000, pp. 93–98).

choices would be linear (and proportional), so that if we double the number of options the time required to choose doubles as well. Why isn't it so?

This is actually easily explained: every time we choose, we do not consider every available option (linear time), but rather we cluster options in categories, dismissing progressively a part of them, and roughly half of the options every time. That also means that for *a* and *b* constants, if the number of options grows, so does reaction time, even if nonproportionally. Vice versa, given an equal number of choices, *a* and *b* influence reaction time.

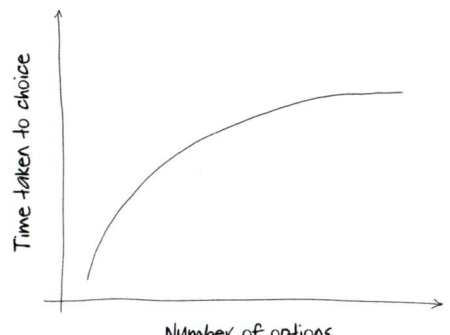

FIGURE 7.7

Hick's law: the ratio between the time necessary to perform a choice and the number of choices is nonproportional.

This is why caution has to be used, for instance, when applying Hick's law to items in a list: if that list does not provide a **meaningful ordering criterion** from the user's point of view, no clustering is possible and the user will probably browse through each item every time. That means that reaction time becomes linear and the formula loses any utility (or—better—the logarithmic curve becomes a straight line, meaning that what was nonlinear becomes linear). However, lists providing some meaningful ordering criterion allow for such scanning and clustering: think of an alphabetically ordered list of animals, where a recognizable principle the user knows how to apply is in force. There is no scanning all of the options, but a quick dive to the pertinent cluster inside the list, say, the letter M for meerkat, and then an evaluation of the relevant subset: all animals whose name begins with M. This time the law applies and reaction time is nonlinear, and you find your meerkat in a flash.

> **Hick's Law** - Hick's law shows that choice is not as much an issue of quantity, of the number of options available, but rather of quality, of the way in which such options are organized and presented to the user.

Remember what we said about **consistency**? Let's go back to Borges's Chinese encyclopedia for a moment, that most peculiar list mentioning animals "belonging to the emperor, embalmed, which from a distance look like flies" and so on. We said that is a canonical example of inconsistency in a classification system. What that means is that in such a system it's impossible to recognize any organization principle. Right. That list is a perfect case in which Hick's law does not apply or where it becomes linear: the reader has no other choice than to browse every single item on the list. We can read it until our eyes hurt, but that list is neither semantic nor alphabetical nor based on any known conventions. Now, as a matter of fact, that list can be considered to be very similar to your run-of-the-mill menu of a typical Web site or to the options list inside an application screen: at times, the only tie that binds these different elements together is so thin or broad that it's basically not there. Thinking in terms of consistency allows a little backtracking: **consistency supports reduction**. When apparently there is no way to operate directly on choices, it might be

Reduction is tightly coupled with consistency

time to see if introduction of a consistent set of options enables more meaningful lists and reduces both effort and anxiety.

That is why at the beginning of this chapter, when recounting Luca's adventures in the supermarket, we said that the problem at hand is not the large number of choices but the way they are organized and presented. That is, the solution is not less grocery, but designing the process in such a way that the scale presents customers with some meaningful organization that aligns or resonates with the whole of the shopping experience. As they are, the numbers on the scales unfortunately only reflect the supermarket's point of view, not the user's, and they are a one-time mechanism disconnected from all other informational sources in the supermarket.[2]

Now, back to the formula: there's one last implication we need to address, but it's going to be quick, we promise. Suppose we have two lists (Figure 7.8): both of them contain eight items, but they are organized differently.

1. List #1 is flat and presents all eight items in one level.
2. List #2 is hierarchical and presents the eight items using two four-item menus. That is, it has two items on the first level, each of them containing four items on the second level.

Suppose as well that the items are organized in a meaningful way: the menus are consistent and Hick's law applies. If we calculate the time required for choosing in both cases using Hick's formula, we see quickly that choosing once from one significantly ordered eight-item menu is quicker than choosing twice from two four-item menus: **wide structures**, with fewer levels, are preferred over deep structures with more levels. If you are ready with your calculator, it goes like this mathematically:

Wide vs. deep structures

1. $a + b \log_2 8 = a + 3b$ in the first case
2. $2(a + b \log_2 4) = 2a + 4b$ in the second case[3]

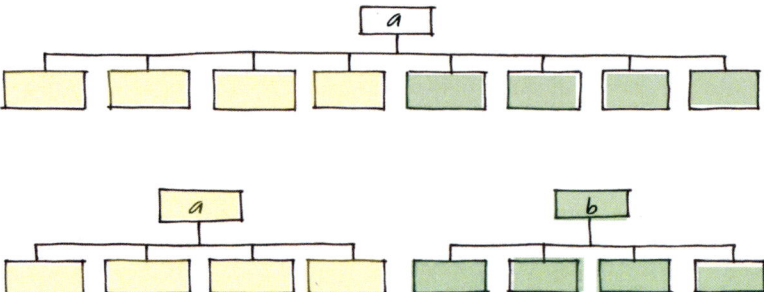

FIGURE 7.8

Two lists, different organizations.

---

[2] Try asking someone from staff if 253 is a local produce, and see if they understand.
[3] Because $\log_2 8 = 3$, while $\log_2 4 = 2$.

and, obviously, $a + 3b < 2a + 4b$, as a and b are positive nonzero constants, established empirically. Their typical values for theoretical calculations are $a = 50$ and $b = 150$. We discuss some of the implications of this more in depth later; for the time being, we just note that Hick's law tells us that reaction time depends not just on the number of choices but also on the way these choices are presented to users. Or: the way by which we organize items within an information space strongly affects the way people interact with them.

REDUCTION IN PERVASIVE INFORMATION ARCHITECTURE

Anderson's economy of niches does not presuppose an infinite global availability of choices for all, but a wider range of options in specific fields—the niches—in which individuals could not previously have them as these were isolated from the larger mainstream markets.

(Dini 2006).

It's more choices then, but only in the niche or domain we are interested in. If we consider Anderson's long tail from the perspective of a specific user, his model actually implies a **reduction in the range of choices**: from all those that are theoretically available to only those that are of some interest. This could be the result of filtering mechanisms that allow discarding unnecessary items; for example, the way the Amazon Web site implements its systems of dynamic correlations: "Frequently bought together," "Customers who bought this item also bought…" "Look for similar items by…" and so forth (Figure 7.9).

Fewer, selected choices at any given moment

The principles of long tail markets and the cognitive triggers being activated when presented with choice actually move in the same direction: building a segment or niche approach to address users' choices transparently in their field of interest might definitely be a valid strategy to **reduce discomfort** and increase satisfaction. Here is what Sheena Iyengar and Mark Lepper[4] write on the subject:

Reducing stressful choices in pervasive information architecture

> despite the detriments associated with choice overload, consumers want choice and they want a lot of it. The benefits that stem from choice, however, come not from the options themselves, but rather from the process of choosing. By allowing choosers to perceive themselves as volitional agents having successfully constructed their preference and ultimate selection outcomes during the choosing task,

---

[4] Iyengar and Lepper are the authors of the study *When Choice Is Demotivating: Can One Desire Too Much of a Good Thing?* quoted by Schwarz in his book. Available at http://www.columbia.edu/~ss957/whenchoice.html.

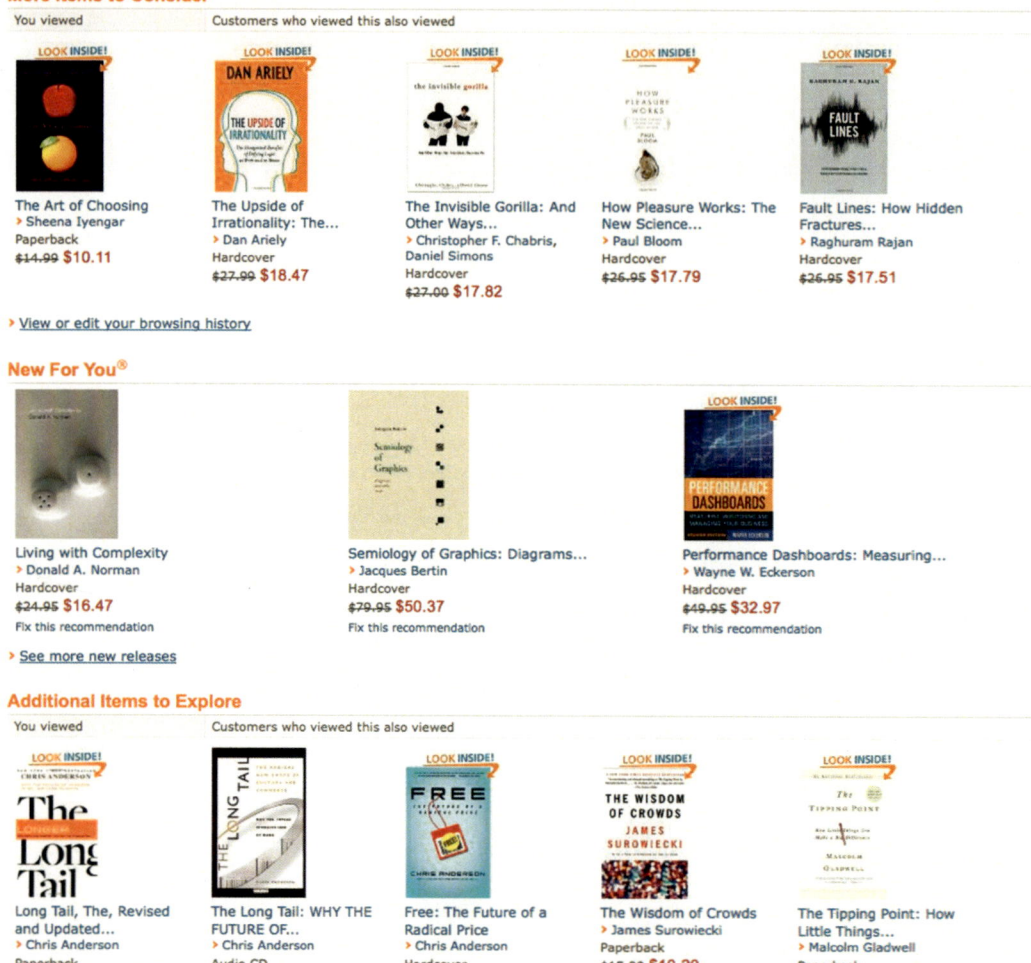

FIGURE 7.9

A few of Amazon's strategies to reduce the paradox of choice.

the importance of choice is reinstated. Consider the request in Forbes' recent "I'm Pro-Choice" article: "Offer customers abundant choices, but also help them search." We now know how.

(Iyengar & Lepper in Anderson 2006, p. 172).

Internal and external reduction

Pervasive information architectures add the usual **internal and external** modifiers to the way reduction operates. Internally, reduction applies to the way we present choice and options in a single channel, for example, the way a generic mobile application does not include the possibility to print at all as it's currently not relevant. We speak instead of external reduction when we consider strategies for reducing choice-induced stress across all different channels similarly: for example, when producing a weekly flyer for a supermarket that separates products on the pages in a way that follows how they are laid out in the store. In this second, wider sense, reduction works together with consistency.

A couple of structural and organizational principles can be deployed practically to increase the reductive capabilities of pervasive information architecture: (1) organize and cluster; (2) focus and magnify.

Organize and Cluster

Luca's electronic do-it-yourself scale in the supermarket is a rather useful example of the way the paradox of choice works. Its display, buttons and icons are a perfect example of a list of options ordered casually or meaninglessly: the numerical sequence is just that, a numerical sequence, and does not entertain any logical relationships with the groceries it supposedly should help weigh and price. Associations between products and numbers change often.

This makes remembering any of these (memory game-like) couplings very difficult. When used, icons often fail to be of any help as they are small, crowded, and distributed randomly.[5] Customers try to level the field by deploying a number of very interesting, creative, but totally unnecessary strategies. Regardless of how we try to cope with the machine, it is evident that this design prevents customers from operating the kind of clustering that Hick's law requires; thus, even if we kind of know the number, reaction time grows linearly and not logarithmically.

To restore the conditions for that to be possible, and hence reduce the time necessary for choosing and the cognitive load that goes with it, we can apply what we call the **organize and cluster** principle.[6] The principle dictates two possible, different practices:

Organize and cluster principle

1. List menu items using meaningful, self-evident rules so that users can cluster items, according to Hick's law.
2. When no ordering is possible, cluster and organize in levels. Nested levels are a possible design strategy because a wide structure offers no advantage over a deep structure if Hick's law does not apply. More than that: the levels themselves restore some degree of clustering.

Not surprisingly, the first solution corresponds to the canon of **helpful sequence** as outlined by Ranganathan, the Indian mathematician and

Helpful sequence principle

---

[5] As opposed to, say, ordering vegetables by color.
[6] Resmini and Rosati (2008, pp. 8–9); Rosati (2007, pp. 67–71).

librarian who invented faceted classification. In his *Prolegomena to Library Classification*, he writes:

> the sequence of the classes in any array should be helpful. It should be according to some convenient principle, and not arbitrary, wherever insistence on one principle does not violate other more important requirements.[7]

(Ranganathan 1967).

When that helpful sequence cannot be discerned or is plain absent, solution number 2 (focus and magnify) helps reintroduce it by splitting the list into shorter, more consistent sublists.

Let's go back once more to the do-it-yourself scale (yes, we like that example). Most supermarkets implement some variant of these, and some stores adopt solutions that improve performance (Figure 7.10).

These scales do not list all available products in one long list, but require an initial choice from a first-level menu that sections the domain in some meaningful way. Some of them, for example, list *Fresh fruit, Vegetables, and Dried fruit*; some just do *Fruit and Vegetables*; and some have a main screen with

FIGURE 7.10

A do-it-yourself scale with submenus: using clustering to reduce the number of simultaneous choices.

---

[7] Some excerpts (including the one quoted) are available in Denton (2009). For an overview, see Denton (2003).

the most commonly bought products and second-level menus for *Fruit and Vegetables*. When customers choose any of these top items, their second-level content becomes visible.

Choosing *Fruit*, for example, would show *Apples, Apricots, Kiwis, Peaches, Pineapples*, and so on. This works much better. Wait. You are objecting: but according to Hick's law, choosing once from an 8-item menu requires less effort than choosing twice from two 4-item menus. Hence, splitting a 60-item or more menu into a 2- or 3-item main menu and a number of subordinate menus should increase the cognitive burden on the customer and fuel the paradox of choice. Isn't that the case here? Well, no. Hick's law does not apply here, as we have an inconsistent list of options. By distributing products onto two levels we certainly increase the number of choices necessary (2 vs. 1), but we simultaneously reduce the number of alternatives customers have to choose from and build more consistent sets.

Focus and Magnify

Contextualization and customization are two other ways to counter the paradox of choice. Amazon's flexible suggestion system is once again a good example. We all know how it works: once we start using their Web site we start receiving in-context notifications. Who bought *a* also bought *b*, if you are interested in *c* maybe you could be interested also in *d*, and so on. What we might not know is that this strategy successfully relies on the basic human attitude to sample and select that Marcia Bates (2002) pinpoints as the founding model of information seeking, a model that in a time span of possibly millions of years exapted from food foraging to information foraging.[8]

We describe this procedure as **focus and magnify**: first you focus on a niche, an item, and then you magnify and look around for similar items in the same area (Figure 7.11).[9] While the end results are analogous to those obtained by applying the organize and cluster principle, focusing and magnifying shift the accent from working on the information side of things to working on the user experience, and it is probably better suited for being applied for internal, single-channel, reduction.

Focus and magnify strategy

In physical environments, this could very well mean that items and products are related in such ways as to allow for adjustments, either by refocusing or magnifying according to the user's seeking behavior; for example, this could be obtained by applying faceted classification techniques and showing the related items/products belonging to the same facet(s) for each facet of an item/product.

---

[8] See Chapter 6, p. 121.
[9] Resmini and Rosati (2008, pp. 8–9); Rosati (2007, pp. 72–74).

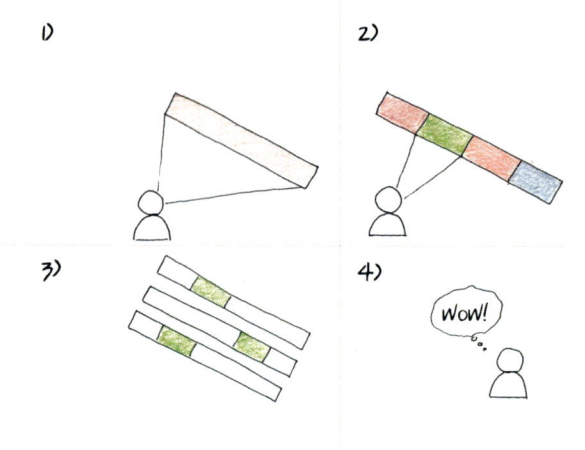

FIGURE 7.11

How to reduce the number of initial choices and enable more relevant results by means of focusing and magnifying.

Take the winery from Chapter 6, for example: different, overlapping informative layers could suggest or highlight wines coming from the same region, those from the same grape(s), or matching the same recipes. Alternatively, a freely faceted classification scheme (Gnoli et al. 2008), tagging, or cross-reference links could be used to the same effect. This could be entirely physical, employing traditional way-finding strategies via labels, signage, colors, and positioning; or it could be digital, via one or more information layers. These would provide, for example, additional, product-based information such as reviews, comments, and comparisons. By tapping into these layers users would be able to save their paths, preferences, and choices for later reelaboration and reuse.

The focus and magnify approach introduces a complementary degree of simplicity and complexity in the information architecture that resonates with the elegant synthesis operated by John Maeda in his *Laws of Simplicity*:

Law 5: Differences

Simplicity and complexity need each other. . . . establishing a feeling of simplicity in design requires making complexity consciously available in some explicit form. . . . The closest approximation to the solution I have found is in the concept of rhythm, which is grounded in the modulation of difference.

Law 6: Context

What lies in the periphery of simplicity is definitely not peripheral. . . . I personally experienced this sensation of being "comfortably lost" on a recent vacation hike in Maine. I noted that trails were marked with rectangles of bright blue paint. Each of the trails was highly navigable due to its good condition, but once in a while I would pause and wonder, "Where do I go next?" And almost like magic one of thew blue markers that previously sat in the background of my perceptual field literally "popped" into the foreground. With my bearings restored, I would slowly return to the beautiful, uninterrupted forest vistas with the emotional satisfaction and comfort that one feels on a mountain hike.

(Maeda 2006, pp. 45–46, 53, 60).

LESSONS LEARNED

Know

- Reduction does not mean taking choice away
  Improving choice by means of reduction does not mean cutting down the number of options available, but it's rather an overall issue of quality in the process, of how these options are logically organized and presented in such a way that users can make the most out of them.

- Having many choices does not run contrary to the long tail model
  The cognitive process of choosing and Anderson's economic model work toward similar outcomes, as the latter does not presuppose an infinite amount of available choices but rather a wider array of options in the specific niches that are of some interest to any given customer.

- Simplicity and complexity are not mutually exclusive
  Complexity is richness: simplicity is a strategy to make this complexity viable, comprehensible. As such they are complementary and positively unbalance the design process.

Do

- Create consistent collections
  Consistency reduces cognitive load and the stress associated with choosing. Enable helpful, meaningful sequences so that users can figure out the underlying logic lists are ordered by.

- Build meaningful structures wide and shallow
  Wide and shallow structures are better than narrow and deep structures. Hick's law demonstrates that choosing one time among a single set of eight ordered options is better than choosing two times among two sets of four options.

- Organize and cluster: go narrow and deep
  If the items in a collection do not lend themselves very well to consistent listing, Hick's law does not apply. Split lists into two or more levels and create smaller clusters with fewer choices to allow clustering.

- Focus and magnify
  Guide users toward their niches as soon as possible and then offer them a wider range of options by using Amazon-like, context-aware horizontal correlations.

CASE STUDIES

The Horizontal Palimpsest

Internet TV. Imagine to choose.   (Sony advertising).

The amount of content from different platforms that finds its way to the TV screen is getting larger and larger: public, generalist channels, in both analogical and digital format; satellite; on-demand IPTV; and streaming Web TV. All of these are either available for free or upon subscription or as pay-per-view one-time deals. The number of TV-enabled devices further adds to complexity: programs can be watched in real time on a TV set, on a smartphone, or via a laptop computer. Programs can be recorded. Video casts can be downloaded from the Web and consumed at leisure on whatever device supports one of the many DIVX or MPEG formats or containers. The same goes for audio content: radios, iPods, MP3 players, mobile phones, computers, you name it. We are beyond multichannel and are moving into cross-channel, where the message has a life of its own and can be reencoded to be broadcast in several different media, with an overwhelming array of choices. Allowing for a consistent user experience across these heterogeneous formats, channels, and devices is mandatory to avoid pitiful user experience and an overall idea of poor branding.

These experiences have to cater to users who look for an active role in a complex context–the users Sterling calls wranglers–by mashing-up, sharing, and (re) producing on one side, and to users who look for simplicity, less involvement, and who are generally more consumer oriented than production oriented on the other.

We could apply the matrix Bates applies to information seeking strategies to this scenario: active vs. passive or directed vs. undirected, with the former being more expensive in terms of time, attention, and competencies and the latter being less expensive and ultimately less engaging.[10] Barry Schwartz (2005, pp. 77–78) calls these two behavioral types *maximizers* and *satisficers* (Table 7.1):

- Maximizers "seek and accept only the best"
- Satisficers "settle for something that is good enough and do not worry about the possibility that there might be something better"

Implementing a coherent organization model for the growing mass of programs, formats, and channels of this contemporary hybrid TV model is no easy task. Providing both active users/wranglers/maximizers and passive

---

[10] See Chapter 6, p. 115.

Table 7.1 Comparison between Behavioral Models from Different Authors in Different Fields

Field	Source	Behavior
Information seeking	Bates (2002)	Passive vs active Undirected vs directed
Technology and design	Sterling (2005)	Users vs wranglers
Cultural and media studies	Jenkins (2006)	Mainstream vs grassroots Multichannel vs transmedia
Social sciences	Schwartz (2005)	Satisficers vs maximizers

users/consumers/satisficers with a good user experience means that we need to manage the way by which choices are presented: on one hand, going through every single option available is less and less a concretely exploitable strategy even for a maximizer; on the other hand, satisficers are no longer completely passive either. We couldn't think of shaping up such a system by means of one single consistent mechanism, in accordance with Hick's law. It is simply impossible: there is too much heterogeneity in both the products and the users' fruition models. This is one perfect paradigmatic example in which—to reduce the paradox of choice or to allow any choice at all—it is necessary to employ the *organize and cluster* and *focus and magnify* strategies explored earlier.

> I was having dinner with a group of friends about a month ago, and one of them was talking about sitting with his four-year-old daughter watching a DVD. And in the middle of the movie . . . she jumps up off the couch and runs around behind the screen. . . . She started rooting around in the cables. And her dad said, "What you doing?" And she stuck her head out from behind the screen and said, "Looking for the mouse."
>
> (Shirky 2008).

The organize and cluster approach suggests the deployment of multidimensional information architectures (such as faceted classification) that allow multiple groupings and pathways into the information set. But this is not enough: interacting with such a large, complex system may still be too stressful and time-consuming even for maximizers.

To lower the bar, we can implement a *focus and magnify* strategy that employs social classification systems. Their contextual, customizable suggestions would help reduce the number of choices to choose from drastically—down from every single option available to the ones users could really be interested in. Once I more or less hit the mark, these focused choices could be expanded by any magnify mechanism. For example:

- Last week I watched *x* and I liked it. I'd love to see something similar.
- I missed the last two shows of serial *y*. I'd love to see them.
- Restart from where I left the last time (a movie or any show)
- I want some adrenaline, I want some romance, I want some mindless fun (goal-oriented)
- I want a movie for a quiet evening, for a romantic evening, for an evening with my group of friends (task-based)

GIANNI BELLISARIO AND CHIARA FERRIGNO—EXPERIENCING TV IN THE AGE OF CROSS-MEDIA

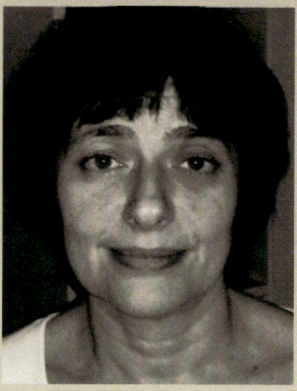

When we talk of television of the future we don't know what that is and how it will be, which appearance and interfaces it will have. It will probably behave and look rather different from what we are accustomed to today, but one thing seems to be pretty clear: it will be all over the place. A survey performed in 2009 on behalf of the Italian Permanent Observatory on Digital Content[11] by Nielsen on Italians between the ages of 19 and 29 shows an evident strong cross-media trend: these young adults mix channels (Web, TV, radio), devices (laptop or desktop computer, smartphone, player), and environments (physical and digital) continuously. They are perpetually jumping across media, with the Web working as a connector.

Traditional broadcasters already produce large quantities of nonlinear programming through a much varied ecosystem of media and channels. Just to name a few of those most well-known: reruns, Web programming, mobile programming, pay TV or pay-per-view offerings, HD services, and catch-up TV.

This in turn has led to the creation of several different vertical, market-specific platforms to manage such a complex set of items and documents, from rights management to macroprogramming scheduling, up to daily broadcasts. However, these tools still do not take into account a vast amount of artifacts that part of the broadcasters offer, such as dedicated Web sites, games and contests, telephone-based interactions (e.g., surveys or voting), merchandising (including DVDs for sale), magazines; user-generated content such as fans communities, and spin-offs (which could provide some hints of whether a product is cross-media friendly).

This is not an exclusive list: the next big thing could be just about to knock at their door. At the same time, a broadcaster's catalog is a valuable asset. This implies that being able to access yesterday's content and information is important. Ordering any such catalog for current and future fruition is

---

[11] Osservatorio Permanente Contenuti Digitali (2009).

GIANNI BELLISARIO AND CHIARA FERRIGNO—EXPERIENCING TV IN THE AGE OF CROSS-MEDIA—CONT'D

no easy task. What we suggest is to flip the table and imagine a different, horizontal palimpsest, fully cross-medial. For broadcasters, organizing and accessing their content and services in terms of this horizontal palimpsest means adopting a radically new perspective. They have been traditionally television-centric: now they are no more.

They have been coarsely producing a generally available product on a day/week/month/year schedule: now they have a much finer granularity they can decline at will, offering the most bizarre content to a few interested individuals, together with the big event everyone wants to see.

This is not cosmetic: this is necessary to move scheduling away from the static verticality of flowing through time only that is the traditional reference grid of broadcasting.

This does not mean incorporating everything in a Big Brotheresque, TV-centric perspective: quite the contrary. When planning cross-media strategies from the outset, it might very well be that another platform is the focal point and that the TV screen is just an afterthought. What it means is resilience and increased choice: giving everyone a personal inroad to get there. For some, this might be software, for example, through a semantic search engine; for others it could be a social network or even a human mediator, a personal assistant. We don't know. We only know we need both approaches to make the ecology work.

Such a system maximizes both the final user experience and the business value of it, as broadcasters better exploit every item in their catalogs. Such a system moves with people across different media and environments and supports their choices as one continuous bridge experience that listens to all of them,

from novice to expert, and provides them with just the right amount of complexity they need.

For example, tomorrow this "thing" could ask us what we want to see: and even if we don't know what to choose, it will choose for us, according to our previous choices, our temporal or behavioral patterns, our social connections. Every morning we will say "the same" to another "thing" (a smartphone, for example) in order to receive our daily menu of videos, news, music or stock quotes, and these two "things" will talk to each other, so we could pick up where we left off regardless of the device we are using. It will be a sophisticated system that keeps it simple for us and that will search the network for us: resources for learning, friends who are in some social networks, products. It will upload or download user generated content, and will keep our personal pages up-to-date.

Chiara works at the cross-media department of Palimpsest Direction of Rai (the Italian public broadcaster). After a career as a screenwriter, Chiara met the Web in 1997: since then she works on multimedia content design. From 2000 to 2002 she was part of a select group of Rai consultants for new technologies; in the following years she has been consultant for Rai TV and multimedia formats.

After long experience in theater (drama and dance production), in 1989 Gianni joined Rai. In 1995 he became director of the new Rai Multimedia department, building the first Internet Web site of the Rai Group, and from 2000 to 2004 the Rai Product Innovation department. Currently, Gianni coordinates the cross-media department of Palimpsest Direction and represents Rai in the Crossmedia EBU (European Broadcasting Union) commission.

These different strategies let users focus on the choices available to them in the areas they are effectively interested in, removing noise—the undesired and unnecessary options—and leaving them with the opportunity to widen their pools to similar or related niches. This effectively allows users to navigate information space along a horizontal axis metonimycally (see Chapter 5) built on similarity and contiguity and allowing much more freedom than a preordered, vertical hierarchy (Figure 7.12 and 7.13). We will discuss this in detail in Chapter 8, Correlation.

FIGURE 7.12

Italian public broadcaster RAI's teletext service: an example of a standard vertical palimpsest that allows only sequential browsing for channels or time.

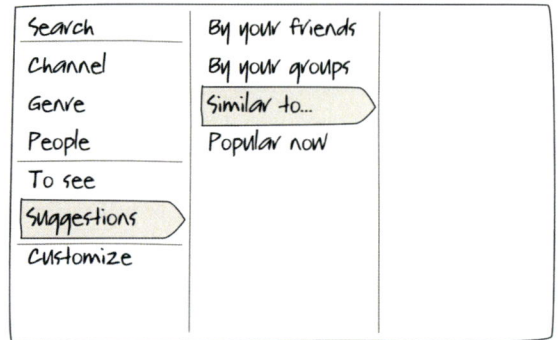

FIGURE 7.13

An idea for a multifaceted horizontal palimpsest that includes custom and social suggestions, allowing for different browsing and searching strategies.

RESOURCES

Articles

Cockburn, A., Gutwin, K., & Greenberg, S. (2007). A Predictive Model of Menu Performance. In *CHI '07 Proceedings of the SIGCHI Conference on Human Factors in Computing Systems*. San Jose, 28 April - 3 May, 2007. ACM Press. http://portal.acm.org/citation.cfm?doid=1240624.1240723.

Iyengar, S. S., & Lepper, M. R. (2000). When Choice Is Demotivating: Can One Desire Too Much of a Good Thing? *Journal of Personality and Social Psychology*, 79, 995–1006. Available at http://www.columbia.edu/~ss957/articles/Choice_is_Demotivating.pdf.

Kuksov, D., & Villas-Boas, J. M. (2010). When more alternatives lead to less choice. *Marketing Science*, 29(3), 507–524.

Lipman, B. L. (2001). Why is language vague? Department of Economics, University of Wisconsin. http://citeseerx.ist.psu.edu/viewdoc/summary?doi=10.1.1.148.2778.

Samuelson, L., & Swinkels, J. (2006). Information, evolution and utility. *Theoretical Economics*, 1, 119–142.

Schmidt, L., Goodman, N. D., Barner, D., & Tenenbaum, J. B. (2009). How tall Is tall? Compositionality, statistics, and gradable adjectives. *Proceedings of the Thirty-First Annual Conference of the Cognitive Science Society*.

Seow, S. C. (2005). Information Theoretic Models of HCI: A Comparison of the Hick-Hyman Law and Fitts' Law. *Human-Computer Interaction*, 20, 315–352. Available at http://citeseerx.ist.psu.edu/viewdoc/summary?doi=10.1.1.86.4509.

Books

Anderson, C. (2006). *The Long Tail: Why the Future of Business Is Selling Less of More*. Hyperion.

Calvino, I. (1993). *Six Memos for the Next Millennium*. Harvard University Press.

Iyengar, S. (2010). *The Art of Choosing*. Twelve.

Norman, K. L. (1991). *The Psychology of Menu Selection: Designing Cognitive Control at the Human/ Computer Interface*. Ablex Publishing. Available at http://www.lap.umd.edu/poms/. See in particular Chapter 8, Depth vs Breadth of Hierarchical Menu Trees.

Raskin, J. (2000). Fitts' Law and Hick's Law. In Raskin. *The Humane Interface: New Directions for Designing Interactive Systems* (pp. 93–98). Addison-Wesley.

Raymond, E. S., & Landley, R. W. (2004). *The Art of Unix Usability*. Pearson Education. Available at http://www.catb.org/~esr/writings/taouu/html/.

Schwartz, B. (2005). *The Paradox of Choice: Why More Is Less*. HarperPerennial.

Wurman, R. S. (2000). *Information Anxiety*. Que.

Videos

Ariely, D. (2008). Keynote. *Authors@Google*. http://www.youtube.com/watch?v=VZv--sm9XXU.

Pixar (2006). Lifted. http://www.youtube.com/watch?v=maR5JEDBltc.

Correlation

LUCA INTRODUCES A GASTRONOMIC INTERLUDE

I'm done with the grocery, and I'm waiting for my turn at the supermarket's meat desk; everyone has a number and the place is rather crowded, as an elegantly dressed gentleman is deep in conversation with the butcher and his assistant for some recipe suggestions. A few persons in the line are getting impatient, as he is spending too much time chatting. But have we met this man before? Wait. Yes! He is the seasoned businessman looking for a wine to go with a green pepper fillet we encountered in Chapter 6. I didn't tell you, but you know something? The winery is just opposite the supermarket, so he bought a couple of bottles of Sagrantino and a bottle of Torgiano Rosso Riserva, and now he's here. Slowing me down. Oh well. I guess I can fill you in while he's busy talking. He will take his time: he still needs the fillet and a few more ingredients and ideas for a few exotic appetizers, a little pasta, some garnish.

Because his office is not far away, our businessman is a regular customer of this specific supermarket, and he is on friendly terms with a number of people from the staff. He usually asks for suggestions when he is looking for something new or when he feels he might benefit from that special pro touch. The supermarket is fairly crowded, and everyone seems busy enough, but he manages to ask a couple of questions while having his fillet cut for him. One of the staff members is from central Italy and has a few interesting ideas: "Why not try a lentil soup? We have a special promotion on the lentils, and there is a selection of some of the finest just at the end of the aisle. If you haven't tried them out, this seems just the right moment. Serve the soup hot with small crostini, well toasted, and add a ride of that extra-virgin olive oil you find next to the lentils when you serve. It's bottled and produced a few kilometers away from where the lentils come from."

This seems like a good idea, so he says his thank yous, much to the relief of us all, and after a couple of minutes spent looking around and reading labels, our friend the businessman decides that his fillet is going to go with lentils from Castelluccio di Norcia, a Protected Geographical Indication product coming from a little town in Umbria, Italy. According to the notes on the package, they seem to match one of the wines he bought – the Sagrantino – perfectly, so the dinner happily takes a central Italy flavor. This gives him ideas. Scrap the appetizers, our businessman thinks, let's make things even tastier by adding a small selection of Tuscan cold cuts for starters. That would be super. He turns around and disappears again among the aisles. And we leave him there. It's my turn and I need some fillet as well.

INTEGRATING THE SOCIAL AND THE INFORMATION LAYERS

Now reload: imagine a different scenario. What if the supermarket had in place some sort of information layer, accessible to customers, where smart tags, thematic paths through the aisles, and links between products were not only exploited to produce a richer shopping experience, but a better global user experience? What if this was not simply plugged into the present, with no memory of events, but it was a more complex system where habits and preferences of individual users were saved for reuse and reconnected for social consumption? Where something like a real-world version of collaborative tagging could be used to supplement and help educated choices. Your friendly staff is still part of the picture, if you need them, but then, not all supermarkets are blessed with competent and customer-oriented personnel, and not every time we have the time to stop and ask. People interactions work synchronously, meaning that a conversation only succeeds if the participants are engaged simultaneously. An information layer can work **asynchronously**.

> **Correlation** - The capability of a pervasive information architecture model to suggest relevant connections among pieces of information, services, and goods to help users achieve explicit goals or stimulate latent needs.

Technology and technologically enhanced human networks are actively increasing the amount of data we produce, receive, process, and transmit: if you remember, we introduced the idea of reusing this bounty of information that mostly lies there in Chapter 6 to improve the resilience of pervasive information architectures. At the same time, these data can be used to improve the relational dimension of the system both in physical and in digital spaces, effectively providing alternative, novel ways to browse, navigate, and discover that are independent from top-down, hard-coded structures. Consider Luca's supermarket in the story and the way the staff created a preferential, all-in-one place for a number of products from central Italy, effectively adding an independent path inside the navigable space of the store for a few temporarily related products (Figure 8.2).

That is nothing new. Supermarkets and shops have been doing this for ages. So, why stop there? Why not apply some of the tricks we learned designing on the Internet, as we said we would back in Chapter 3? Take it one step further and make these suggestions part of a consistent, resilient, information architecture that supports customers, where users are an integral part of the human–information interaction process (Marchionini 2004), where customers are coproducers, wranglers, and remediators. This system sustains the social, collaborative patterns of social networking and applies them to the environment and the objects within (Sterling 2005). User-generated and object-generated data are used to propose alternative classification or exploration models; multiple social categorizations of products (top views, other users also bought, reviews); collaborative tagging of physical items are used to improve their findability.

Just remember: this is a design problem, not a technology problem. There are strings attached. For example, as American experience architect Joe Lamantia wrote in one article for the online magazine *UX Matters*,

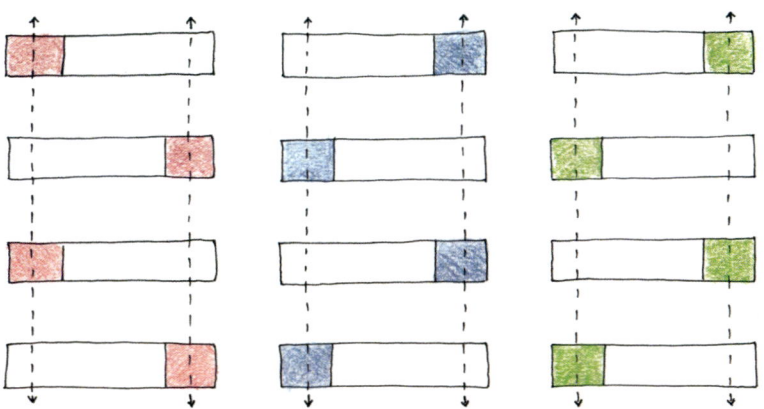

FIGURE 8.2

Correlation creates novel paths independent of top-down, hard-coded structures in both the digital and the physical environment.

the social maturity of current augmented experiences is similar to that of a young child who is learning the complex rules and norms that determine socially acceptable behavior. With unevenly developed abilities and understanding, fitting into social situations is very difficult.

(Lamantia 2010).

Lamantia mentions augmented experiences, that is, real-time, computer-mediated experiences where sound or video is embedded into physical reality but this can be generalized easily to ubiquitous ecologies, and it is not difficult to see that this might seem extreme to some of us, some sort of a (possibly) benign, marketing-driven dystopian future from a Philip Dick novel. But we urge you not to allow yourself to be stopped by any prejudice: we are going there. Maybe it does not look as if we are going fast, but we are gaining ground steadily. Even though privacy and the protection of personal, sensitive information are everyone's most important concern, powerful traction is being exerted by all those who have an urgent need to control: paternalistic governments, evil corporations, family control freaks, prospective burglars, or plain lunatics. We all have our favorite examples of someone we wouldn't want to know what we are doing.

Hiding in a closet and wishing this away is not going to work. Poking fun of technophiles does not work either. Sometimes the shades of gray vastly outnumber any other color, and this is one of those cases. Think of the old story of turning your fridge into an Internet-aware appliance: we heard Adam Greenfield sort of ridiculing the idea at the ASIS&T European Information Architecture Summit in Amsterdam, The Netherlands, in 2008. Greenfield was keynoting, and we are all suckers for a good joke during a keynote, but the reality of things is slightly different. As much as an Internet-aware fridge might seem silly to many of us, what about elderly people who need home assistance or people with disabilities? Wouldn't it be great if the person in charge of the groceries could receive a list of what is running short of supply without being there physically? They could buy all that is necessary en route and then visit. What about those of us who chronically forget the milk or the cheese?

Then there is a second aspect: not everything can be foreseen, not all uses can be predicted right at the start. Nobody thought the invention of the telephone would revolutionize human communication: spending hours chit-chatting? Never. What for? Isn't visiting more proper? The telephone was simply a way to accelerate business mail, to the point that many of the early installations connected individual companies to their banks. In this sense, correlation is also linking unrelated sources together for unexpected results: people, information, objects, and correlation can be one powerful wheel when it comes to generating complexity. If you let us go grab a glass of water, we have a couple of examples we can share with you.

THE CASE OF THE BROAD STREET PUMP

"No data yet," he answered. "It is a capital mistake to theorize before you have all the evidence. It biases the judgment."

(Conan Doyle 1995).

FIGURE 8.3
The streets of London in Roman Polanski's *Oliver Twist* from Charles Dickens's novel by the same name (2005).

It is 1854. London is a sprawling, ever-growing urban agglomerate of roughly 2.5 million people, ready to overtake Paris both in size and in importance. Its populace has grown rapidly and inordinately, almost tripling the number of houses in the city between 1801 and 1851. Queen Victoria is on the throne, a duty she will honor until her death in 1901, and this is Victorian London, the London we know from such novels as Charles Dickens's *Bleak House*—its streets teeming with people, carriages, and horses, its markets vociferous, its stench potent:

> the homes of the upper and middle class exist in close proximity to areas of unbelievable poverty and filth. Rich and poor alike are thrown together in the crowded city streets. Street sweepers attempt to keep the streets clean of manure, the result of thousands of horse-drawn vehicles. The city's thousands of chimney pots are belching coal smoke, resulting in soot which seems to settle everywhere. In many parts of the city raw sewage flows in gutters that empty into the Thames. . . . Personal cleanliness is not a big priority, nor is clean laundry. In close, crowded rooms the smell of unwashed bodies is stifling. It is unbearably hot by the fire, numbingly cold away from it. At night the major streets are lit with feeble gas lamps. Side and secondary streets may not be lit at all and link bearers are hired to guide the traveler to his destination.

(Perdue 2010).

With such an explosive growth, human waste in the city has become a huge problem, one surprisingly aggravated by the recent diffusion of the flush toilet, a new technical marvel that unfortunately empties sewage straight into the Thames, and by what we could call issues of early globalization.[1]

As a result, all throughout the first half of the 19th century, London was repeatedly hit by Asiatic cholera epidemics. Cholera is a diarrheal fever with a mortality rate, if untreated, which still reaches today a scary 50% of the infected. At the time, the fact that the infecting agent was not known did not help. A first episode in 1832 caused more than 5200 deaths. A second one, ravaging London in the years 1848–1849, killed nearly 14,800 (Figure 8.4).

On August 31, 1854, a new cholera outbreak struck Soho. In the space of 3 days, 127 people died, all of them living either on Broad Street, now Broadwick Street, or in its immediate vicinity. By September 10, the number of victims had risen to roughly 500 and most of the residents of the neighborhood had fled their houses. In all, this epidemic would kill more than 11,600 Londoners. As Judith Summers (1991) writes in her history of Soho, the neighborhood was

FIGURE 8.4

Illustration from the *Punch* magazine (1852). *Source: Wkimedia.*

A COURT FOR KING CHOLERA.

---

[1] Namely, the recent discovery of the virtues of South American guano, bird droppings, as a fertilizer. The introduction of guano meant that night soil men, professionals who emptied cesspools, started having a hard time reselling their main asset to farmers and quickly resolved that discharging human waste in the rivers, Thames included, was the way to go.

an insanitary place of cow-sheds, animal droppings, slaughterhouses, grease-boiling dens and primitive, decaying sewers. And underneath the floorboards of the overcrowded cellars lurked something even worse—a fetid sea of cesspits as old as the houses, and many of which had never been drained.

The Broad Street pump was one of many water pumps that the streets of cities such as London had at the time. Houses did not have running water, and pumps were a fact of life for drinking, for cooking, and for the little hygiene that could be managed in popular districts. The Broad Street pump was, of course, on Broad Street, at the intersection with Cambridge, and many people were dying around there. Still no one considered water to be a problem. A doctor, John Snow, now considered one of the fathers of modern epidemiology, was proposing at the time a somewhat unconventional notion that the disease did not spread by means of miasmatic bad air, but by ingesting small particles. He had been trying to prove this theory for some time, but because this was long before Pasteur and microbes, neither the authorities nor his colleagues were really paying him attention. The epidemic in Soho provided him with the ability to gather sufficient data to maintain, scientifically, that cholera was contracted by drinking sewage-contaminated water.

What is most interesting to us is how much of Snow's relentless inquiry was circumstantial, experimental, and not medical in nature. His methods often could be described today as user centered. For example, one brewery just a few yards away from the pump suffered no loss of lives among its employees. Snow investigated and found out that they were given free beer as part of their wage and hence never drank the water from the pump.

Some of the victims did not seem to be linked to the Broad Street pump, the most famous of whom was a widow living in Hampstead. Both she and her niece had died. Snow went all the way to her house and found out that the widow had lived on Broad Street once and that she loved the taste of the water from the street pump so much she had sent a servant to bring some back every day, the last time on August 31. He interviewed people. He went to see how and if their cesspools were overflowing or leaking. He took samples of water from the pump.[2] Above all, Snow was incredibly good at connecting apparently unrelated facts and sources into his vision and had a knack for making it visual: his original map detailing the number of victims and their location in respect to the pump is still incredibly informative (Figure 8.5).

In the end he managed to get the authorities to agree, albeit reluctantly, to remove the pump handle "as an experiment." The outbreak, which had already slowed down in the days before, came to a stop.

---

[2] And found it to contain "white, flocculent particles."

FIGURE 8.5
Detail of the original map with cholera cases (in black) drawn by Dr. John Snow in 1854 (Broad Street highlighted in red by the authors). *Source: John Snow Archive and Research Companion.*

John Snow was applying a fresh mind to a well-known unsolved problem. He connected loose facts into a new, coherent, and scientific view capable of explaining the otherwise unexplainable. Breaking down silos—avoiding that important knowledge remains confined in specialized containers—is probably the major take away from the case of the Broad Street pump after more than a 150 years from the facts. Unfortunately, we have not paid that much attention so far: the channels we use today to communicate are still very often sealed, self-contained, autoreferential boxes. We know another doctor who can confirm this and cheer us up a bit while he's at it.

BREAKING DOWN THE SILOS

At the 3rd Italian Information Architecture Summit in Forlì, in February 2009, one of the speakers—medical doctor Gian Piero Perscarmona (2009)—brought an instructive (and hilarious) case study on the difficulties of establishing a correct diagnosis for a patient, a case that could be easily made into a *Dr. House* screenplay.

Official medical channels have no place for user forums, patient mailing lists, or postsurgery online communities. However, the Web has a billion Web sites where people discuss their health issues, share impromptu solutions to everyday illnesses, and comment on the various results and side effects that medications or hospitalization have on them. Pescarmona explained that for a professional physician, this **user-generated knowledge** is a gold mine, void of conditioning from both years of praxis and the longa manus of the pharmaceutical industry.[3] He proceeded to explain where the gold is with an example.

Social medicine

He was passed on a particular patient, a gentleman in his mid-thirties, rather chubby, extremely calm, by a colleague. The diagnosis said hyperthyroidism. He had his blood samples examined, and the intricate interplay of values and parameters seemed to support this conclusion. Pescarmona mentioned how he has developed this insane idea that silos are bad things, so he wasn't satisfied and he wasn't thinking inside the box: something was not right. People suffering from hyperthyroidism are usually thin, stretched, and edgy—nothing like the person he had in the chair in front of him.

He made another appointment for a week later and began a hunt on the Web. He found plenty of references of people suffering from the same symptoms, and they were all related to medical use or abuse of psychotropic substances, such as anesthetics and analgesics. Thus, at the next meeting, he cautiously started to inquire, and finally asked his patient almost off-hand "Do you by any chance use any antidepressants? Anxiolytics?" The gentleman replied as if almost shocked: "Sleeping pills? Who? Me? Never, not a single medication." He was clearly disgusted: "But I sure do cocaine."

Correlation, the idea of networking heterogeneous resources in a larger, more meaningful mesh that enables new solutions to emerge and creates new opportunities to exploit, seems to be overseen too easily, which is rather bizarre, considering how much information spaces depend on that most promiscuous artifact, the hyperlink. Luckily we have brilliant people the likes of Andrew Hinton to remind us.

---

[3] He also added that this view of his is seen as rather eccentric and a tad too radical by many of his colleagues.

ANDREW HINTON—LINKS, MAPS, AND HABITATS

It's strange how, over time, some things that were once rare and wondrous can become commonplace and practically unnoticed, even though they have as much or more power as they ever had. Consider things like these: fire, the lever, the wheel, antibiotics, irrigation, agriculture, the semiconductor, the book. Ironically, it's their inestimable value that causes these inventions to be absorbed into culture so thoroughly that they become part of the fabric of societies adopting them, where their power is taken for granted.

Add to that list two more items, one very old and one very new: the map and the hyperlink. Those of us who are surrounded by inexpensive maps tend to think of them as banal, everyday objects—a commoditized utility. And the popular conception of mapmaking is that of an antiquated, tedious craft, like book binding or working a letterpress—something one would only do as a hobby, since, after all, the whole globe has been mapped by satellites at this point and we can generate all manner of maps for free from the Internet.

But the ubiquity of maps also shows us how powerful they remain. And the ease with which we can take them for granted belies the depth of skill, talent, and dedicated focus it takes for maps (and even mapping software and devices) to be designed and maintained. It's easy to scoff at cartography as a has-been discipline—until you're trying to get somewhere, or understand a new place, and the map is poorly made.

Consider as well the hyperlink. A much younger invention than the map, the hyperlink was invented in the mid-1960s. For years it was a rare creature living only in technology laboratories, until around 1987 when it was moderately popularized in Apple's HyperCard application. Even then, it was something used mainly by hobbyists and educators and a few interactive-fiction authors, a niche technology. But when Tim Berners-Lee placed that tiny creature in the world-wide substrate of the Internet, it bloomed into the most powerful cultural engine in human history.

And yet, within only a handful of years, people began taking the hyperlink for granted, as if it had always been around. Even now, among the digital classes, mention of "the web" is often met with a sniff of derision. "Oh that old thing—that's so 1999." And, "the web is obsolete—what matters now are mobile devices, augmented reality, apps and touch interfaces." One has to ask, however, what good would any of the apps, mobile devices, and augmented reality be without digital links?

Where these well-meaning people go wrong is to assume that the hyperlink is just a homely little clickable bit of text in a browser. The browser is an effective medium for hyperlinked experience, but it's only one of many. The hyperlink is more than just a clicked bit of text in a browser window—it's a core element for the digital dimension; it's the mechanism that empowers regular people to point across time and space and suddenly be in a new place and to create links that point the way for others as well.

Once people have this ability, they absorb it into their lives. They assume it will be available to them like roads, language, or air. They become so used to having it, they forget they're using it—even when dazzled by their shiny new mobile devices, augmented reality software, and touch-screen interfaces. They forget that the central, driving force that makes those technologies most meaningful is how they enable connections—to stories, knowledge, family, friends. And those connections are all, essentially, hyperlinks: pointers to other places in cyberspace; they are links between conversations and those conversing—links anybody can create for anybody to use.

This ability is now so ubiquitous, it's virtually invisible. The interface is visible, the device is tangible, but the links and the teeming, semantic latticeworks they create are just short

ANDREW HINTON—LINKS, MAPS, AND HABITATS—CONT'D

of corporeal. Like gravity, we can see its physical effects, but not the force itself. And yet these systems of links—these architectures of information—are now central to daily life. Communities rely on them to constructively channel member activity. Businesses trust systems of links to connect their customers with products and their business partners with processes. People depend on them for the most mundane tasks, such as checking the weather, to the most important, such as learning about a life-changing diagnosis.

In fact, the hyperlink and the map have a lot in common. They both describe territories and point the way through them. They both present information that enables exploration and discovery. But there is a crucial difference: maps describe a separate reality, whereas hyperlinks create the very territory they describe.

Each link is a new path, and a collection of paths is a new geography. The meaningful connections we create between ourselves and the things in our lives were once merely spoken words, static text, or thoughts sloshing around in our heads. Now they're structural—instantiated as part of a digital infrastructure that's increasingly interwoven with our physical lives. When you add an old friend on a social network, you create a link unlike any link you would have made by merely sending a letter or calling them on the phone. It's a new path from the place that represents your friend to the place that represents you. Two islands that were once related only in stories and memories are now connected by a bridge.

Think of how you use a photograph. Until recently, it was something you'd frame and display on a shelf, carry in your wallet, or keep stored in a closet. But online you can upload that photo where it has its own unique location. By creating the place, you create the ability to link to it—and the links create paths, which add to the ever-expanding geography of cyberspace.

Another important difference between hyperlinks and traditional maps is that digital space allows us to create maps with conditional logic. We can create rules that cause a place to respond to, interact with, and be rearranged by its inhabitants. A blog can allow links to add comments or have them turned off; a store can allow product links to rearrange themselves on shelves in response to the shopper's area of interest; and a phone app can add a link to your physical location or not at the flick of a settings switch. These are architectural structures for informational mediums; the machinery that enables everyday activity in the living Web of cyberspace.

The great challenge of information architecture is to design mechanisms that have deep implications for human experience using a raw material no one can see except in its effects; it's to create living, jointed, functioning frameworks out of something as disembodied as language and yet create places suitable for very real, physical purposes. Information architecture uses maps and paths to create livable habitats in the air around us, folded into our daily lives—a new geography somehow separate, yet inseparable, from what came before.

Andrew Hinton is a principal user experience architect at Macquarium, a UX consulting firm headquartered in Atlanta, Georgia. An internationally recognized speaker and writer on IA and UX, Andrew has designed information systems and interfaces for Fortune 500s, small businesses, and nonprofits alike. He continues to be involved with local and international practitioner communities, such as IA Institute, IxDA and UX Meetups and Book Clubs. Andrew is a big believer in the practice of information architecture, which (to his mind) concerns the design of contexts and their connections in digital space. Andrew lives in Charlotte, North Carolina, blogs at inkblurt. com, and tweets via @inkblurt.

AT THE HAWTHORNE GRILL

There's this passage I got memorized. Ezekiel 25:17. . . . I been saying that shit for years. And if you heard it, that meant your ass.

(Tarantino 1994).

Quentin Tarantino's award-winning movie *Pulp Fiction* tells a rather common, been-there-done-that story of mobsters in modern drug-dealing Los Angeles, but its pace, narrative flow, and cinematic language are nothing usual. The movie breaks up the story and its continuity in so many ways that we can recompose the full picture only when the credits are rolling. And not easily, by all means. In an essay written in 2000, Fiona A. Villella says that *Pulp Fiction*

> stops and starts, shifts and rewinds, forcing the viewer to construct the story—the trajectory of each character, their interrelation with other characters and fictions, the "how," "what," "when" and "why" of the narrative. *Pulp Fiction* has a circular narrative. At certain moments where the narratives intersect, the theme of the uncanny and destiny arises, for example, where Butch and Vincent pass each other at Marsellus' bar. They exchange hostile glances and comments for no apparent reason. The sequence is mysterious, and Vincent's immediate reaction of hostility toward Butch proceeds unexplained. Of course, later on, in the story concerned with Butch and his escape from the LA mob, he comes across Vincent and kills him.

Pulp Fiction is actually three separate episodes, tightly interwoven: *Vincent Vega and Marsellus Wallace's Wife*, *The Bonnie Situation*, and *The Gold Watch*. These stories are broken up into scenes that are then shuffled around like pieces in a puzzle, with chronologically contiguous events moved away from one another, with no respect for the "correct" timeline. *Pulp Fiction* nonlinear storytelling is not even really circular, but more of a Mobius strip, with the Prologue and the Epilogue at the Hawthorne Grill welded together.

Now, a Mobius strip is a curious object and a neat party trick: by twisting one of the ends 180 degrees when you close the ring you end up with just one surface instead of two. Give it a try, and be puzzled: this is exactly how you feel when the lights go on after you see the movie for the first time. These two sequences at the Grill—where Pumpkin and Honey Bunny (played by Tim Roth and Amanda Plummer) are first discussing the job and then trying to deliver while Vince and Jules (the two hit men played by John Travolta and Samuel L. Jackson) are having breakfast—are basically the same interrupted scene twisted around a little bit and shown from different points of view (Figures 8.6 and 8.7).

FIGURE 8.6
Pulp Fiction, Q. Tarantino (1994).

FIGURE 8.7
Circular storytelling in *Pulp Fiction*: the Prologue and the Epilogue sequences at the Hawthorne Grill create a narrative Mobius strip.

So, *Pulp Fiction* is really a series of small pieces loosely joined, a collection of narrative elements that slowly and controversially add momentum and meaning to each other way past the moment when all the stories are complete. But it is also a movie loaded with citations from the most diverse and unexpected sources: TV, cinema, music, and consumer products. Citations are all around. Some of them pop, for example, Bruce Willis's character Butch the boxer considering a chainsaw as a possible weapon in the style of Sam Raimi's *Texas Chainsaw Massacre*; some of them quite sophisticated, such as Tarantino's introducing John Travolta and Uma Thurman's famous dance scene as an homage to French cult director Jean-Luc Godard's surreal musical moments (his movie *Bande à part*, 1964, comes to mind). Two elements of the movie are strategic from our architectural point of view:

- circularity: how the story ends (or does not end) where it started
- recombination: how narrative pieces are moved around and connected

These are the base characteristics that enable correlation, and you know what? They are not new at all.

While the movie has been somewhat responsible for giving recombinative narrative, or *entrelacement* as it is called in linguistics, a new exciting visibility,[4] the technique in itself has been around for ages. Addressing Tarantino-type narratives with pensive remarks connecting them to postmodernism only gets close to the mark. Certainly citationism, intertextuality, refusing a central single point of view are typical of postmodern culture, and Tarantino's nervous camera work and incredibly chatty action owe a great deal to the **stylemes** of modern American pop culture.

> **Styleme** - Any recognizable and repeatable trait of a director's style that effectively contributes to his or her signature, a unique visual style. It might be the use of color, cuts, movements of the camera, or a combination of these. The word *styleme* was introduced in the mid-1990s by Peter Wollen as analogous to the concept of a phoneme in linguistics and is now often applied to many of the visual arts, including comics and cartoons.

---

[4] Movies using this alternate cut technique as a central element of their narrative include such diverse examples as Terry Gilliam's *12 Monkeys*, David Lynch's *Lost Highway*, and Christopher Nolan's *Memento*. Hard-core Stephen King's fans and fantasy aficionados will remember also how the theme of circularity, the wheel of Ka, is central to the gargantuan saga of the *Dark Tower*.

But entrelacement? Not new, to the point that one of the best examples of this technique can be found in one of Italy's major romantic epics from the Renaissance, the *Orlando Furioso*, the Orlando enraged, or, as it is commonly known in the English-speaking world, the *Frenzy of Orlando*.

THE FRENZY OF ORLANDO

But fierce Ferrau, bewildered in the wood,/Found himself once again where late he stood.

(Ariosto, Orlando Furioso).[5]

Doujinshi - This is particularly huge, for example, in Japan, where the doujinshi aniparo phenomenon has generated both a thriving market and a number of very successful meetings and conferences, but it is far from being neglectable everywhere, as *The Hunt for Gollum* fan movie connected to Peter Jackson's *Lord of the Rings* trilogy can testify (http://www.thehuntforgollum.com).

Ludovico Ariosto wrote his masterpiece between 1510 and 1532 as an explicit continuation of Matteo Maria Boiardo's *Orlando in Love*, very much like aficionados from all over the world reprise their favorite heroes in follow-up **fan-made stories**. While the historical backdrop goes back to the times when Charlemagne and his paladins were fighting the invading Saracens, Ariosto has very little consideration for verisimilitude, space–time continuity, or historical accuracy. His story is again actually three different stories weaved into 48 cantos and more than 38,000 lines, all of them mixed up with magic, fantastic creatures, and incredible voyages:

1. Orlando's search for Angelica
2. The love adventures of Bradamante and Ruggiero
3. The war between Christians and Saracens

Italian literary critic Leonzio Pampaloni (1971) has compared the narrative of the Furioso to a Lego building made up of red, white, and black bricks. It's a good metaphor, and just to stretch it a little bit further, we could analyze the text taking down the building and making three ordered piles, one for each color. Or, more interestingly, we could try to understand how they have been connected by means of entrelacement or interlace.

Entrelacement works like cinematographic cuts. Here are some examples: "Return we now to him, to whom the mail/Of hawberk, shield, and helm, were small protection:/I speak of Pinabel the Maganzeze" (Canto 3, IV) or "Leave we sometime the wretch who, while he layed/Snares for another,

---

[5] Quoted verses from Canto 1, XXIII. *Orlando Furioso* is available on Project Gutenberg in the translation of William Stewart Rose (http://www.gutenberg.org/ebooks/615).

wrought his proper doom;/And turn we to the damsel he betrayed,/Who had nigh found at once her death and tomb" (Canto 3, VI).

Whatever method we choose, we will find out quickly that the *Frenzy's* apparent chaos is really an incredibly fine-tuned mechanism, where the madness of Orlando works as a central element that divides the entire epic in two symmetrical sections (Figure 8.8) and where the various ligatures either separate closely related elements or correlate totally separate ones.

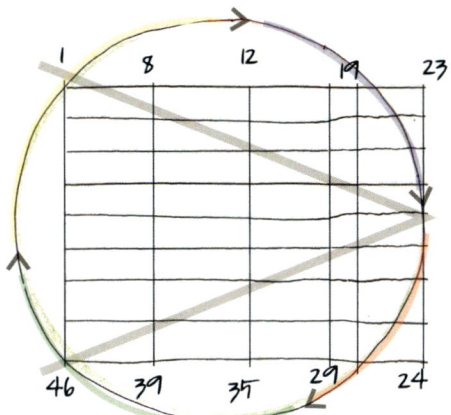

FIGURE 8.8
Entrelacement in the *Frenzy*: the madness of Orlando neatly divides the work in two symmetrical sections. Numbers indicate the cantos; the apex of Orlando's madness is between cantos 23 and 24.

METANARRATIVE

He breaks off narrations verie abruptly so as indeed a loose inattentive reader will hardly carrie away any part of the storie.[6]

On July 4, 1969, at the Festival dei Due Mondi in Spoleto (Italy), one of the largest theatrical events in Italy and Europe, Luca Ronconi debuted his version of the *Frenzy of Orlando*. Edoardo Sanguineti, writer, poet, and translator, was the playwright. This is what the Italian researcher and director Claudio Longhi (2006, p. 7) writes in his book recalling the creative process that led to that theatrical adaptation:

> the Orlando furioso lies on the green moquette in director Luca Ronconi's study. Ludovico Ariosto's octets, transformed into dialogs by Edoardo Sanguineti, are glued to large sheets: every paper rectangle hosts four simultaneous scenes corresponding to the four different themes into which, in its theatrical rendition, the many adventures crowding the epic will be split.

Ronconi's approach to the *Frenzy* was unique. While Ariosto interlaces the three narrative threads and breaks down their original unity, Ronconi decided to play a different, opposite game. With the help of Sanguineti, he unraveled the tangle and recomposed the integrity of the story, adding one little variation for theatrical purposes. He split the poem into four themes instead of three: Orlando's search for Angelica; Ruggero and Bradamante's love adventures; the war between Christians and Saracens; and the madness of Orlando. What they did then is absolutely fascinating.

It went like this: Ariosto chose three largely unrelated threads and used entrelacement to weave them into one winding, circling narrative, breaking

---

[6] John Harington thus wrote in the preface to his own translation of the *Orlando Furioso* in 1591.

FIGURE 8.9

The *Frenzy of Orlando*, Luca Ronconi. Audience members and cast mix up without any distinctions between what's play and what's real. *Photo: P. Manzari, long-time actor of the Orlando.*

up their original unity. Ronconi and Sanguineti, some five centuries later, take a good look at this complex creation and restore all threads to their original state (and while they are at it, they even identify a fourth one). Well, you say, so what? Remember that short quote from Longhi given earlier? Each paper sheet held four simultaneous scenes, each corresponding to one of the four themes in the play. That's exactly what they did on stage, or better, the stages: the different narratives unfolded simultaneously, played out by actors. Events that were somewhat linked narratively, stylistically, or thematically played out at the same time in different areas of the theatrical space.

Ronconi's stage left behind all ideas of a separation between what is fiction and what is real: the audience had no chairs to sit on and was forced to follow the events as a crowd would do, standing, sharing a rectangular area of roughly 18 per 25 meters with the actors. Theatrical props were maneuvered in full sight, and what to follow and when to follow were totally up to the individual spectator. It was like being immersed in events happening in real time (Figure 8.9).

CORRELATION IN PERVASIVE INFORMATION ARCHITECTURE

So, what is all this rambling about correlation? Why should correlating different elements be central when designing pervasive information architectures? And, most of all, why have we been talking about such diverse things as *Pulp Fiction*, the *Frenzy of Orlando*, and a cholera outburst ravaging London more than a 150 years ago?

The reason is simply explained: entrelacement, usually badly executed, is what we normally experience when dealing with today's cross-channel strategies for services or platforms.

The experiences we have with services are no less intertwined than the adventures of Ferrau in the *Frenzy of Orlando* or than *Pulp Fiction*. We are simply less conscious of that: we are so used to considering this the normality of things, jumping back and forth among channels and environments, that it does not really hit us. For us to take notice, for us to see the gaps, something really has to go spectacularly wrong. However, it is so easy to see how correlation between these distributed scampers of user experience is not really an add-on feature but a necessity that we complain all the time when we encounter the walls of some information or experiential silo that should not be there. How to achieve this goal then? How do we break down the silos?

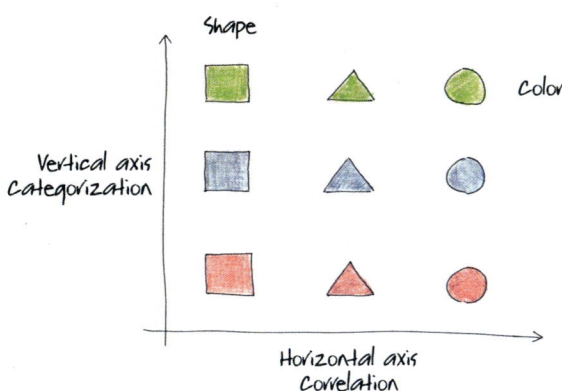

FIGURE 8.10
The two axes of pervasive information architectures.

First of all, this idea of pervasive information architecture design as the design of human–information processes spanning multiple channels and bridging the physical and the digital implies, in turn, a significant shift in the notion of what takes precedence in terms of organization. Generally, classic information architecture seems to be all about taxonomies and hierarchies. Even when these are overtly subverted (think of folksonomies), they still linger around as some obtrusive guest you cannot really get rid of. Correlation, instead, is meant to introduce into the design process a **second axis** that emphasizes the value of horizontal relationships among items: coordination, similarity, and semantic links (Figure 8.10), force us to reconsider our perspective.

Furthermore, and you will not be surprised at this point, correlation is actually two different things as well:

- internal correlation, which promotes semantic proximity between similar items belonging to the same channel
- external correlation, which promotes semantic proximity between items belonging to different channels but connected to the same task, process, or people

Think of the complex entrelacement that the various threads weave in the *Frenzy*, or in *Pulp Fiction*. Those are two beautiful examples of **internal correlation**.

Then consider the two different versions of the *Frenzy*, the text written by Ariosto in the 16th century and the theatrical rendition by Ronconi in the 1960s. Do not focus on them as individual work of arts, but consider the relationships they explicitly

Two axes - In every pervasive information architecture, two axes or dimensions exist: a vertical axis, representing the hierarchical relationships between the items in the collection; and a horizontal axis, representing the similarity links between those same items. Correlating means empowering the traversal, horizontal dimension of information architecture over the vertical one. For more on this, see Chapter 9.

Correlation and the other heuristics - Correlation strategies of course impact on other heuristics. Correlation helps reduce the paradox of choice (reduction, Chapter 7, especially when dealing with focus and magnification), supplies alternative and custom navigation paths (resilience), and ultimately facilitates a berry-picking approach (place-making, resilience).

or implicitly have with each other and with many other texts and plays, such as Boiardo's *Orlando in Love* or other stage works by Ronconi. The way they relate and open up new, unanticipated paths that suddenly enlarge our view are an example of external correlation. Pervasive information architectures especially thrive on this second, cross-channel flavor of correlation.[7]

One more example. Consider once again the supermarket from our opening story. The basic idea behind having part of a shelf (or a visible, well-defined spot) dedicated especially to the week's specials is to promote a system of suggestions and relationships very akin to "if you like this item you may also like" or "who bought this also bought." If we take this one step further and make it into a digital layer integral to the store's information architecture and capable of preserving histories, we can add more exploratory suggestions, can try to anticipate or predict choices, and render latent needs explicit in a way that is uniquely individual (you bought pasta sauce, you will need that pasta you usually buy on weekends). When we connect the two systems, we have external correlation working with internal correlation to produce a complex system.

In a generic business scenario, this translates, for example, to the possibility for users to:

- start a task in any of the channels comprising the ubiquitous ecology of the company and seamlessly complete it in another one—for instance, placing an order on the phone, receiving updates via the Web site, and picking up the purchase at the store (Figure 8.11)
- retrieve and exploit pieces of information acquired and results of tasks performed in channel *a* inside the company's ubiquitous ecology when we move to another channel *b*
- experience unbroken flow along any of the channels or touch points by effectively making them communicating, bridge artifacts (Figure 8.12)

As we wrote when introducing resilience in pervasive information architectures, users are not passive consumers anymore. The manifesto maintains that users are intermediaries, actively shaping and reshaping information

---

[7] It might be interesting to note that correlation has a way of impacting on consistency and vice versa. If you think of Ronconi's *Frenzy*, you quickly come to the conclusion that to narrate a story successfully that is correlated to another being experienced through a different channel (external correlation), such as Ariosto's *Frenzy*, external consistency can be somewhat broken or lessened. This is a constant pattern that you can also apply to place-making for example (see Chapter 4). We discuss this in more detail in Chapter 9.

FIGURE 8.11

External correlation allows for logical and experiential continuity across all channels.

space, codesigners of these new ubiquitous ecologies. Their implicit choices and behaviors generate complexity and meaning as much as their explicit, intentional actions. This conversely means that tasks they perform, behavioral patterns they can exploit successfully or unsuccessfully, and events occurring at any moment along the process influence, either positively or negatively, the holistic experience they have. This also retroactively impacts the user's perception of single touch points in the system: bad global experiences, bad touch points. That's how we define *information circularity*. Designing correlation in pervasive information architectures means enabling such circularity to work to reinforce and not hinder the final user experience.

JONAS SÖDERSTRÖM—THREE CIRCLES

There are three very different kinds of "design" that go into any digital system and they have a certain correspondence to specific users' behavior.

The first, and most obvious, is the visual design. Let's call it *graphic design* for now—form and color and all the visual building blocks, such as lines, tints, type, white space, pictures, and shading. These project mood: this looks nice and modern, that looks serious and trustworthy, or this looks busy and urgent, but that looks dull, that is not serious. It's a quick, effortless, almost subconscious impact. And it's important because first impressions often last.

Then a user will typically try to find or do what she is after—her purpose for the visit. Where's news about China? Where's the form for travel expenses? How do I buy a ticket?

This is the stage where typically the user reads the menus, checks the headlines, looks at links, and clicks the icons. She tries to find her relevant actionable area. She tries to grasp the system's organization to see which path leads her to her goal. If graphic design has a lot to do with "to feel" and (hopefully) "to like," this step involves instead a lot of "to interpret," "read," or "understand."

We could call this "information design." It involves word skills, such as finding understandable names for the subsections and the right words for the navigation menus. Writing clear links and

headlines, but also organizing the content into chunks that make sense, prioritizing them, and finding the best place for them—on the page, on the site, or on the application screens. Also, putting the right metadata in place, to make things findable.

On the designer's side, this clearly calls for different skills. It's not graphic design. The graphic designer will typically use words such as *space* and *alignment* when choosing his design. He seldom has much interest in which piece of information, based on user needs, should be higher up on the page, as long as it is "balanced."

In a properly designed system, the user will find her content, her place of destination, smoothly and directly and will complete what she set out to do. Sometimes this might be simply to read something. Other times there's a stronger element of "to do" or "to make" in it, such as download the expenses form, upload a photo, pay for a pair of shoes, register as a new user, or log in as a returning one. Even if it's "just to read," there will probably be an action element connected to the clicking necessary to move from page to page.

In this stage, many new things have to be designed. And again, it's not graphic design (what does the menu bar look like?) and it's not information design (what words in the menu?). Now we have to decide what happens with the menu when you click on it.

to do, to make, to act

interaction design

graphic design

information design

to see, to feel   to find, to understand, to interpret

FIGURE 8.14

JONAS SÖDERSTRÖM—THREE CIRCLES—CONT'D

Interaction design is a good word for this. Its building blocks are controls, buttons, forms and fields, menus, and their behavior. Can the user go back a step or pause? When she's finished, does she get clear feedback? If something goes wrong, do we provide useful error messages?

We globally call these three stages the information architecture of the system, and this framework for analyzing the design in terms of what the user experiences has proved very successful over several years, and not only for the Web. Imagine an answering service. First, the design of sound: tone of voice, speed, warmth, pitch. Second, its organization: how many choices? How many levels? ("Press one for . . ."). Third, the interaction: can I press 1 before the phrase is finished? Can I go back or not? See?

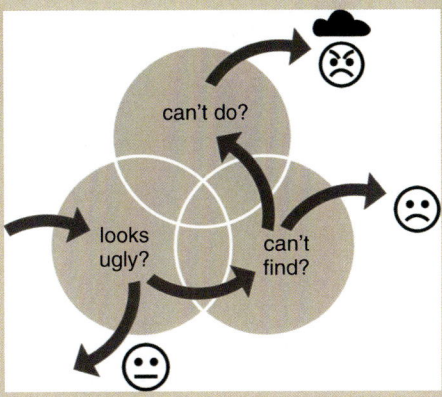

FIGURE 8.15

The framework can also explain how the user reacts to her experience with the system. If she doesn't approve of the visual, she might leave. But probably not too upset. If she can't find what she's looking for, she will eventually give up and leave—significantly more frustrated ("It should be here somewhere"). But if she has found what she needs, then tries to download it, or buys or registers and fails, anger could be a better description of her feelings. "I've spent 10 minutes filling out this form and suddenly it erased everything!"

So graphic design is a threshold to acceptance. Information design comes next and is where probably the most common problems out there lie. Bad interaction design has the greatest potential to harm your brand for a very long time because the user experiences it last and after some significant effort to move through the two initial stages. These three moments are different, but when a system goes wrong, it's almost always because cooperation between them fails and the user is not accompanied from stage to stage.

Jonas Söderström is one of Sweden's pioneer information architects. He has worked with the Swedish Government and the Swedish Parliament and with companies such as IKEA and SonyEricsson. In 2010, he published Jävla skitsystem! (Stupid bloody system!), with insights on how badly designed corporate systems create stress in the workplace.

LESSONS LEARNED

Know

- Correlation breaks down silos
  Correlation creates paths and possibilities and therefore creates shared meaning from isolated, sometimes otherwise useless pieces of information.
- Correlation creates cross-channel continuity and discovery
  Places are palimpsests where people write and rewrite their interactions with the environment, with other people, and with objects. Correlation connects interlaced environments, people, and objects and provides continuity and discovery across channels.

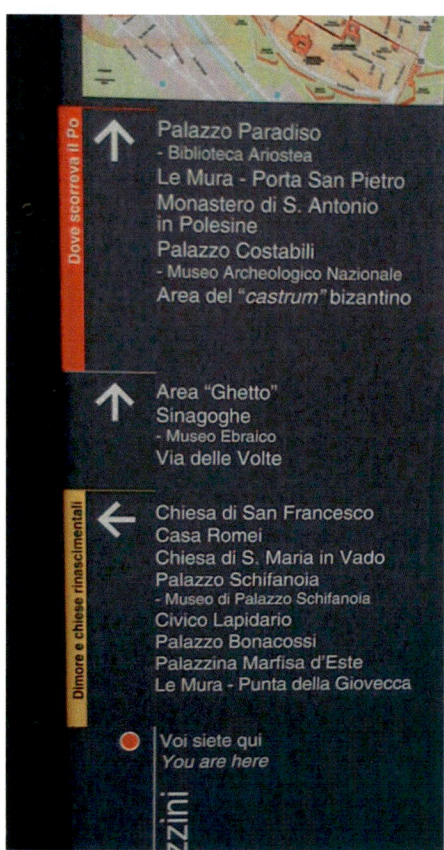

A

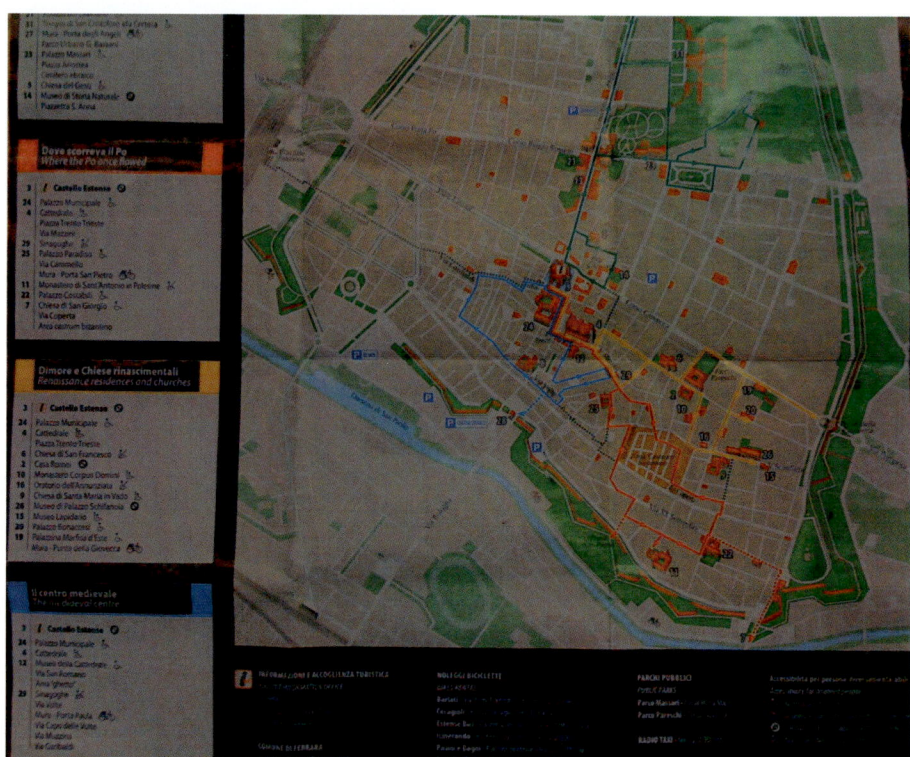

B

FIGURE 8.12

A bridge experience built through a simple correlation of colors, labels, graphics, and language: touristic maps and fixed signs in Ferrara, Italy.

- Correlation can be either internal or external
  Internal correlation links resources pertaining to the same channel, whereas external correlation, which is prominent in pervasive information architectures, correlates resources across channels

Do

- Empower the horizontal axis
  Do not focus only on the hierarchical relationships between items (parent–child, part of a class, etc.): strengthen horizontal relationships such as those implied by similarity, coupling, or social behavior
- Support serendipity and discovery
  Use correlation to elicit unexpressed needs by means of unexpected or not-so-obvious connections
- Exploit both internal and external correlation
  Break down the silos: connect items across channels and do not limit your information flow to one channel at a time

CASE STUDIES

Customer Care

Correlation can be a way to solve that gap between the design of external services (those applied to one's products), and the design of internal services (those applied to noncustomer facing processes, services, and to the in-house organization of the company). We had a hands-on experience of what the lack of very simple correlation means when finding out that a set of earphones we bought from the Apple Web site was not working.

> Day 1: we call Apple customer care to have it replaced. The operator reassures us that it will be replaced immediately via express courier delivery. Day 2: the package is delivered as promised. Unfortunately, it only contains the connection cable and not the earphones. We call Apple customer care to explain that something didn't work out as expected. But this new operator cannot find the previous request for replacement, even though we have an e-mail with a ticket number in it. She then tries with the product serial number, but that fails as the system does not recognize the code. After half an hour on the phone and after explaining the problem to a number of different people at different levels in the Apple customer care organization, our operator explains to us that the system, which at this point has become our unfriendly gatekeeper, has issues with accepting certain serial numbers. Again, we are reassured that, this little problem notwithstanding, our ticket has been passed on to Tech Assistance, Level 2, and that either these people or the courier will get in touch with us.

FIGURE 8.13

Unplugged information, channels, or company departments work like silos. Photo: G. White. *Source: Flickr.*

Day 3: nothing happens.

Day 4: we call Apple customer care again. In a whirlwind of jingles, Level 1 offloads us to Level 2 and Level 2 moves us to Yet Another Level. Here a sympathetic operator understands our befuddlement, has us wait and hold for a few minutes, and then announces to us she has solved the problem.

Day 5: we receive the complete earphone set. The courier knows nothing about sending our nonfunctional set back though.

Day 6: we receive another complete earphone set. Again, no idea of how or if we should send the old earphones back. We guess that if Apple needs them, they will ask.

As the operator we talked to on day 4 explained to us, every single time we spoke to a different person belonging to a different level, a new file had been opened—all of them concerning our single set of earphones. None of them connected. None of them was aware of each other, even if they evidently shared our earphones' unique serial number. The various departmental information systems do not communicate (Figure 8.13): they do not know what is happening throughout the whole customer care process (Figure 8.14). What was for us a single frustrating experience, "talking to Apple," was for them a series of isolated calls from someone claiming he had already called more than a few times.

This is far from being an exceptional case. It is actually the rule for most services and a primary reason for added frustration in our interactions with them.

But what happens when instead of just being short of earphones we are dealing with healthcare, road security, or other services whose prompt answers might make all the difference between safety or risking someone's life?

Patrick Lambe, whom we already mentioned in Chapter 5, has collected reports of various incidents where the stolid preservation of information silos and the lack of any correlation among sources, departments, and entities resulted in tragic outcomes. According to Lambe, some of the most common issues can be ascribed to

- a culture of not caring about the implications of knowledge held beyond a narrow task-fulfillment role
- different ways of describing and naming the same problem
- inability to integrate multiple perspectives on the same problem
- incompatible information systems

- few shared attentional cues among the parties involved—warnings from external parties are not taken seriously because there are no mechanisms for recognizing their authority or the experience upon which the warnings are based (Lambe 2007, p. 53)

RESOURCES

Articles

Dictionary of Victorian London. http://www.victorianlondon.org.

Feldman, S. (2004). The High Cost of Not Finding Information. *KM World*, March 1, http://www.kmworld.com/Articles/ReadArticle.aspx?ArticleID=9534.

Lamantia, J. (2010). Playing Well with Others: Design Principles for Social Augmented Experiences. *UX Matters*, March 8. http://www.uxmatters.com/mt/archives/2010/03/playing-well-with-others-design-principles-for-social-augmented-experiences.php.

Perdue, D. (2010). Dickens' London. *David Perdue's Charles Dickens Page.* http://charlesdickenspage.com/dickens_london.html.

The John Snow Archive and Research Companion. http://johnsnow.matrix.msu.edu.

Villella, A. F. (2000). Circular Narratives: Highlights of Popular Cinema in the '90s. *Senses of Cinema, 3.* http://archive.sensesofcinema.com/contents/00/3/circular.html.

Books

Barabasi, H. (2003). *Linked.* Plume.

Conan Doyle, A. (1995). *A Study in Scarlet.* http://www.gutenberg.org/ebooks/244.

Hempel, S. (2007). *The Strange Case of the Broad Street Pump: John Snow and the Mystery of Cholera.* University of California Press.

Koch, T. (2005). *Cartographies of Disease. Maps, Mapping, and Medicine.* ESRI.

Robertson, R. (2009). *Mock-Epic Poetry from Pope to Heine.* Oxford University Press.

Spence, R. (2001). *Information Visualization.* Pearson.

Summers, J. (1991). *Soho: A History of London's Most Colourful Neighborhood.* Bloomsbury Pub. Ltd.

Vinten-Johansen, P., Brody, H., Paneth, N., Rachman, S., & Russell Rip, M. (2003). *Cholera, Chloroform and the Science of Medicine: A Life of John Snow.* Oxford University Press.

White, J. (2007). *London in the 19th Century.* Vintage Books.

Wollen, P. (1997). *Signs and Meaning in the Cinema.* BFI Publishing.

Movies

Bouchard, C. (2008). *The Hunt for Gollum.* http://www.thehuntforgollum.com.

Polanski, R. (2005). *Oliver Twist.*

Tarantino, Q. (1994). *Pulp Fiction.*

PART

3

Synthesis

CHAPTER 9

Designing Cross-channel User Experiences

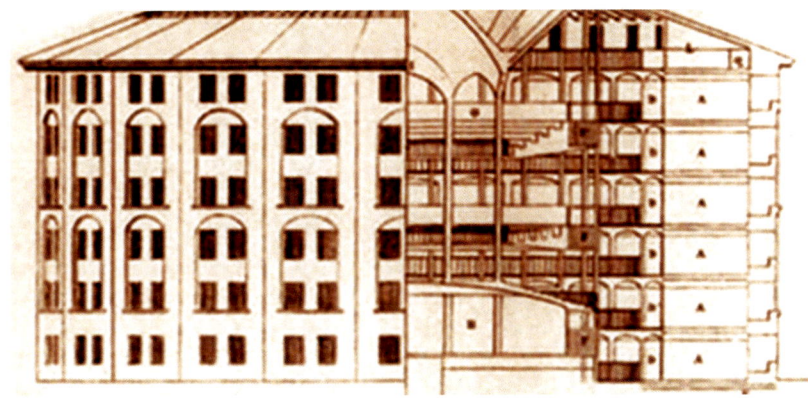

FIGURE 9.1
J. Bentham, Panopticon.

THE TWO DIMENSIONS OF INFORMATION ARCHITECTURE

When we information architects think of the design of an information space, we usually think first of some kind of taxonomy or tree, focusing attention on the parent–child relationships between a set of primary items that we identify as constituting the skeleton. However, alongside this *vertical* dimension there is the complementary *horizontal* dimension discussed in Chapter 8 (Figure 9.2). This axis is of extreme importance in ubiquitous ecologies and is concerned with the way two or more items, despite belonging to different or vertically distant categories, present a logic–semantic correlation (Rosati 2007) capable of tying them together regardless of the channel they happen to be part of. Nonetheless, these links and relationships are more difficult to assess and certainly less structured than those that can be found along the vertical axis; this horizontal dimension is the one where most of the magic of user-generated innovation and unpredictability happens. This is where we really go berry-picking.

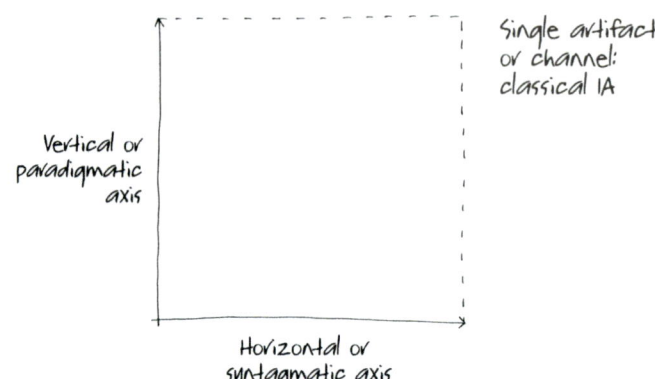

FIGURE 9.2

The two dimensions of information architecture.

The idea of these two orthogonal axes is not anything we came up with: it originates in the work of Swiss linguist Ferdinand de Saussure and is largely used in linguistics and semiotics where they are called the *paradigmatic* and *syntagmatic* **axes**. The model has been widely applied to a number of different languages and contexts, including, for example, cinema and video games, and as we said we think it is a useful way to look at information spaces (Figure 9.2). On the *vertical* or paradigmatic axis lie all hierarchical relationships that each and every item belonging to an information space has with each and every other item; on the *horizontal* or syntagmatic axis lie all the semantic and contiguity relationships that each and every item has with all other items, irrespective of their physical or logical collocation in space, time, or categories.

Paradigmatic and syntagmatic - Linguistics and semiotics refer to what we call the vertical and horizontal axes as *paradigmatic* and *syntagmatic* relations: the former expressing the relationship between any item occurring in a sentence and all the other items that might take its place in that same sentence; the latter expressing the relationship between any item in the sentence and the contiguous ones by which this is variously influenced. For instance, in the sentence "*I'm buying a piece of furniture for my office,*" "*piece of furniture*" has a paradigmatic relation to terms such as *chair, armchair, whiteboard, table,* and so on. These could all be used in the phrase instead of *furniture.* In that same sentence, instead, "I," "am," and "buying" are connected via a syntagmatic relation, as each of those items requires the other: the verbal form present continuous requires *to be + -ing*; the first person "I" requires "am" and vice versa: you cannot say "*I are buying*" or "*I am buy,*" at least if you want to follow the rules.

One important characteristic of the two axes for pervasive information architectures is that they work differently in respect to the internal–external dialectic we highlighted when introducing heuristics in Chapter 3. The vertical axis is mostly an in-channel structure: it describes relationships that are entirely connected to one single channel at a time. Even when you have vertical constructions appearing in different channels (say, IKEA's categorization for furniture), it is just mirrors, very much like working with symlinks or shortcuts on a computer's file system. They are copies or soft copies, not really items being shared between one vertical axes pertaining to different channels.

The horizontal axis is different: in pervasive information architectures it is both an in-channel structure and a cross-channel structure. This is new: in the traditional view of digital artifact design, horizontal axis is implicitly limited to items belonging to the same channel: a Web site, a result set from a search, or data belonging to an application. And while it's of course true that so far we have seen great results from it just being applied as is (say, user-generated correlation on the Web), the real potential is in the way it can link artifacts, people, and information across an entire ubiquitous ecology.

This is the key concept behind the idea of pervasive information architecture. If *place-making* is essential to making people stay and feel comfortable, and *consistency*, *resilience*, and *reduction* help people make sense of what they have around, *correlation* is the backbone of the horizontal axis, the one that conceptually makes the process one single, flowing layer.

BEYOND FLATLAND

I call our world Flatland, not because we call it so, but to make its nature clearer to you. . . . In such a country, you will perceive at once that it is impossible that there should be anything of what you call a "solid" kind.

You see you do not even know what Space is. You think it is of Two Dimensions only; but I have come to announce to you a Third—height, breadth, and length.[1]

(Abbott 1995, section 1, 16).

We started out a couple hundred pages back trying to compare how Jesse James Garrett's much-loved and respected user experience model from the early 2000s would fare if it was to be used to design these new hybrid and dynamic ubiquitous ecologies. We found out we needed to address issues that Garrett wasn't considering at the time. We introduced a different way to think about information architecture. What now? Where are we supposed to take this thing to see how it works? What have we learned?

Well, for once we can see the change now. We can clearly see that, in the design of cross-channel user experiences, information architecture is the diffuse, pervasive, ever-present layer that holds all the pieces together. It is a radical change that positions pervasive information architectures differently from what classical information architecture did. It is not (only) about labels, taxonomies, or menus. It is not about Web sites. It is about design and working

---

[1] Jess McMullin and Samantha Starmer (2010) titled their 11th ASIS&T IA Summit presentation "Leaving Flatland." Although we came to the book from different paths, it seems just fair that we acknowledge their incredibly interesting work with a wink and a nod.

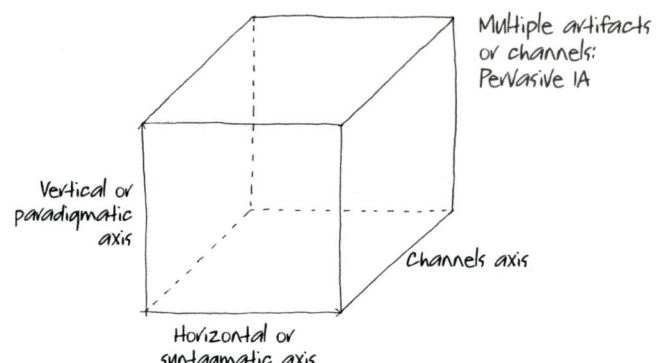

Multiple artifacts
or channels:
Pervasive IA

Vertical or
paradigmatic
axis

Channels axis

Horizontal or
syntagmatic axis

FIGURE 9.3
From two-dimensional
to three-dimensional
information architecture.

with information as the raw material with which we can shape meaning and purpose in more than one domain at a time.

And we can clearly see a pattern. Now that the whole set of heuristics has been deconstructed and explained we acknowledge this horizontal/vertical duality that we tried to pin down when documenting how you could always apply any single heuristic either internally (vertically) or externally (horizontally) as a key characteristic structuring the process.

Pervasive information architectures thrive on this tension between what works inside the silo of a single channel and what works at the cross-channel, ecology level. This is what moves them from the two-dimensional landscape of Flatland to a three-dimensional new world of connected possibilities, where everything acquires volume and thickness (Figure 9.3).

> Behold this multitude of moveable square cards. See, I put one on another, not, as you supposed, Northward of the other, but on the other. Now a second, now a third. See, I am building up a Solid by a multitude of Squares parallel to one another. Now the Solid is complete, being as high as it is long and broad, and we call it a Cube.
>
> (Abbott 1995, section 19).

INTO THE FOURTH DIMENSION

We human beings are time-bound entities. So are all our creations. We cannot think, analyze, measure, prove, disprove, hypothesize, argue . . . without a flow of TIME through our flesh. So we are not objects, but processes. Our names are not nouns, but verbs.

(Sterling 2005, p. 53).

The inhabitants of Flatland (Figure 9.4) cannot perceive a world in 3D "because (they) have no eye." We have no such problem: this is why we can see very well that it certainly does not end there. There definitely is a "Fourth Dimension, which my Lord perceives with the inner eye of thought" (Abbott 1995) and of course this dimension is time.

Pervasive information architectures are dynamic systems, and they change through time: how to account for the actions of active users, Sterling's wranglers, that interact with shared pervasive information architecture through time, and how in turn that architecture changes in response and proposes new uses, new paths, this is what *resilience* mostly addresses in the framework. But all heuristics are concerned with this fourth dimension: the architectures they help build are seamless processes, and there is ample interplay between them.

For example, consider *place-making*: the process through which users consolidate their (hi)stories in the spaces they inhabit, thus transforming them into places, happens through time. It is not there at the beginning. For the sake of an easy, single-channel example, Twitter (or Facebook) was an empty house before users started to add content, connections, and complexity.

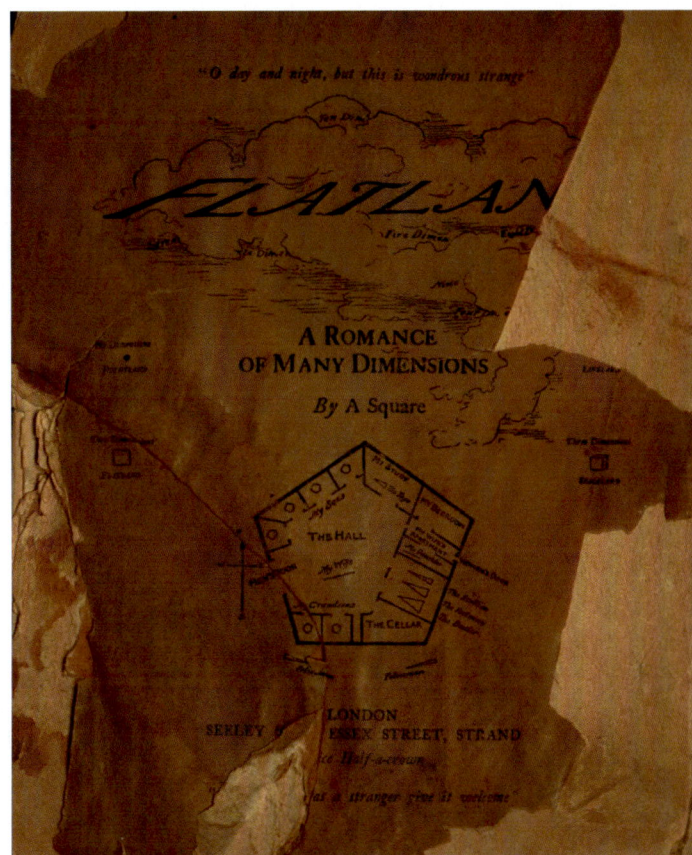

FIGURE 9.4
The cover of the first edition of *Flatland. Source: The Internet Archive.*

The fact that these forays into designing information through time happen in cross-channel systems may lead to interesting, largely unexplored repercussions: Anne Friedberg—scholar of cinema and media studies and author of the beautiful book *The Virtual Window: From Alberti to Microsoft*—states that convergence and contaminations between media not only modify our perception of space, but—inevitably—our perception of time. The most striking example of this phenomenon is an extended continuity between the past and the present (Figure 9.5) and their simultaneous availability. This might not seem that much to anyone who is used to having the Internet around, and places such as YouTube constantly making anything visible and accessible, but it is definitely a big deal if you grew up in the 1970s and early 1980s (or early on, of course).

FIGURE 9.5

Historypin, an application that overlays Google Street View imagery with historical photos, is an early and interesting example of time-flattening and present–past convergence. *Screenshot from Historypin.com.*

Those generations experienced a sudden loss when their go-to movies were shelved and their favorite TV and radio shows went off the air: these were lost, gone, cherished only in memories. Then, suddenly, after more than 20 years, faces, voices, and places straight from childhood or their teenage years were there again for the taking: anyone this side of the Atlantic with even a passing interest in series such as *The Persuaders* knows what we are talking about.[2] As Friedberg says, images are moving from private to public consumption, blending space and time into our own variant of a cyberspace-enabled a eternal "now" (Figure 9.6):

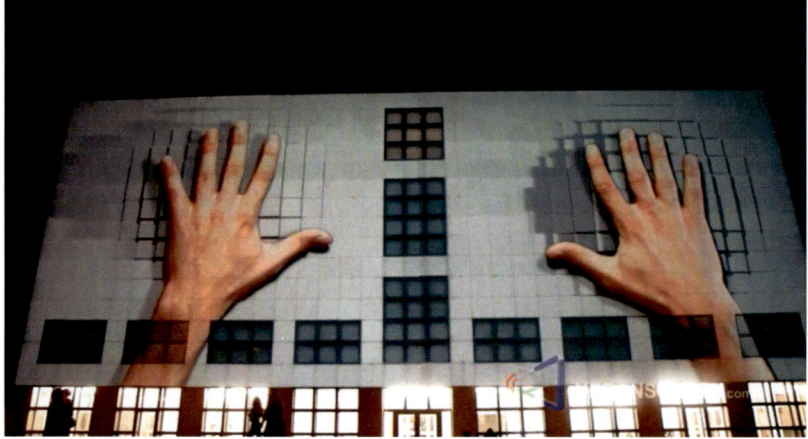

FIGURE 9.6

"555 Kubik," a performance by Urban Screen, is a beautiful example of physical–digital convergence. *Screenshot from Urbanscreen.com.*

---

[2] And here is the obligatory link to the pilot, featuring the famous score by John Barry: http://www.youtube.com/watch?v=xckIh7C4LYg.

as public building and domestic spaces boast image-bearing glass skins, as large-screen television are big enough and fat enough to substitute for real windows, as "windows" within our computer screens stream images from multiple sources, as virtual reality technologies expand from the gaming world into entertainment or daily services, the "virtual window" has become a ubiquitous portal—a "wormhole"—of pasts and futures. . . . As films like *ExistenZ*, *The Matrix*, and *Strange Days* predict, the screen may dissolve.

(Friedberg 2006, pp. 242, 244).

BRINGING IT ALL BACK HOME

We have come full circle. We started out saying with Andrew Hinton that the real breakthrough was the networked hyperlink, and now the hyperlink has taken us out of Flatland.

So, let us be clear: there is no single useful way to describe how to go about the nuts and bolts of actually building a ubiquitous ecology. There is no *how to design a pervasive information architecture in six easy steps*.

As said before, this book is about design and is not that concerned with promoting any specific methodology. We are not going to hand you an easy recipe to follow: two wireframes, a few flowcharts, three iterations, a pinch of salt, and off you go. This would be unfair and wrong. Instead, we give you the full set of Lego bricks we have been playing with for quite a few years now: because they are the basic set, they come in bright primary colors and have simple aggregation rules, and you can build amazing things with them. We just explained to you why: how and what are entirely up to you.

But being 3D is a difficult, unstable balancing act: being able to visualize how exactly the interplay of the different elements works out would bring clarity and a deeper degree of confidence to the table.

One thing we can do, then, is help you understand what basic elements *we* use when we work on pervasive information architectures and how we try to visualize this Spaceland.

First of all, vertical/horizontal duality is something we always try to keep in front of us. We sketch the axes, print out one of the examples from the Web site, and stick a reminder to our wall or whiteboard. We have them handy and often highlight the horizontal axis with a pink marker, as that's the important one. Once we do this, the axes are taken care of. Our motto is just stare at them at least 5 minutes every day, or until you feel dizzy or weak in the knee.

Then we have the heuristics. We use those as our petri dishes. We measure design elements against the rules of thumb they introduce. But what design elements? Well, the channels for sure. Those depend on your specific project, of course, but the Web, mobile platforms, the printed paper, the radio, the phone, and the physical environment are all examples of channels your pervasive information architecture may or may not be concerned with. And then we have one more element: user tasks, both macro and micro. After all, users and their actions are one of the important pieces of these dynamic systems and the touch points where something is bound to happen. Makes sense.

These are the building block of what we call the CHU model, from Channels, Heuristics, and User tasks. Everything fine so far? Good, because it gets a little more complicated than that. We guess that if you paid attention you probably have something in the back of your mind now, nagging restlessly: out of Flatland, in a 3D world, we have three indicators. That means three axes. Hmm. Now that is an interesting problem.

When we presented rough initial sketches of this book to friends and colleagues, to wear them off and have them say "yes, that's incredibly brilliant," we used to show them a swim lane–like diagram that illustrated, visually and with rather snazzy colors, how the CHU elements were impacting on each other.

Figure 9.7 represents one of these swim lanes, detailing the various tasks connected to getting some kind of physical at the local hospital. You can see the user tasks top, then the channels (then labeled environments or media), and finally the heuristics. It was an okay diagram. It allowed us to make our elements concrete and explain, if imprecisely, our ideas. But then, that is still Flatland, isn't it? We needed something that could visualize and convey in one single image the layering and multidimensionality. That's why we drew the CHU cube.[3]

FIGURE 9.7

Heuristics, User tasks, and Channels in a flat swim lane diagram.

Tasks	Prescription or request	Accessing #1 Parking	Accessing #2 Finding the Unit	Hospital #1 Moving around	Hospital #2 Finding the right room / service	Hospital #3 Related tasks
Environments or media		BUILDINGS	BUILDINGS	BUILDINGS	BUILDINGS	BUILDINGS
		SIGNAGE	SIGNAGE	SIGNAGE	SIGNAGE	
			TOTEM	TOTEM		TOTEM
	PAPER					PAPER
		GPS				
	WEB					
	SMARTPHONE	SMARTPHONE	SMARTPHONE	SMARTPHONE	SMARTPHONE	SMARTPHONE
Place-making						
Consistency						
Resilience						
Reduction						
Correlation						

---

[3] Also affectionately known to us as the CHUbe, especially on those days when we spend way too much time fiddling around with it. And of course it is usually nothing like a cube but more of a parallelepiped, as user tasks normally vastly exceed the channels that the pervasive information architecture being designed participates in.

The CHU cube is a 3D view over design space (Figure 9.8). On its three axes lie the CHU indicators: channels are on the *X* axis, heuristics on the *Y* axis, and user tasks on the *Z* axis. The heuristics themselves are usually listed in the way you have seen them presented in the book, with *place-making* at the bottom of the pile, as that's where we start from, but there is no particular reason for preserving this order other than some logical progression from building up to opening up.

The cube is a simplified view, of course: it does not take into account any modification through time, for example. It is just a better rendition of the way the various heuristics, user tasks, and channels interact in a generic pervasive information architecture at a given project time.

We usually draw it with its *Y* layers (the heuristics) flattened paper thin so that we can visualize them properly, but the CHU cube is actually some kind of an information architecture Rubik's cube. The small colored squares shown in Figure 9.8 where user tasks, heuristics, and channels overlap can be considered small colored cubes and they represent how a certain heuristic *h* acts on user task *u* in channel *c*. If a specific colored square is not there, it means that

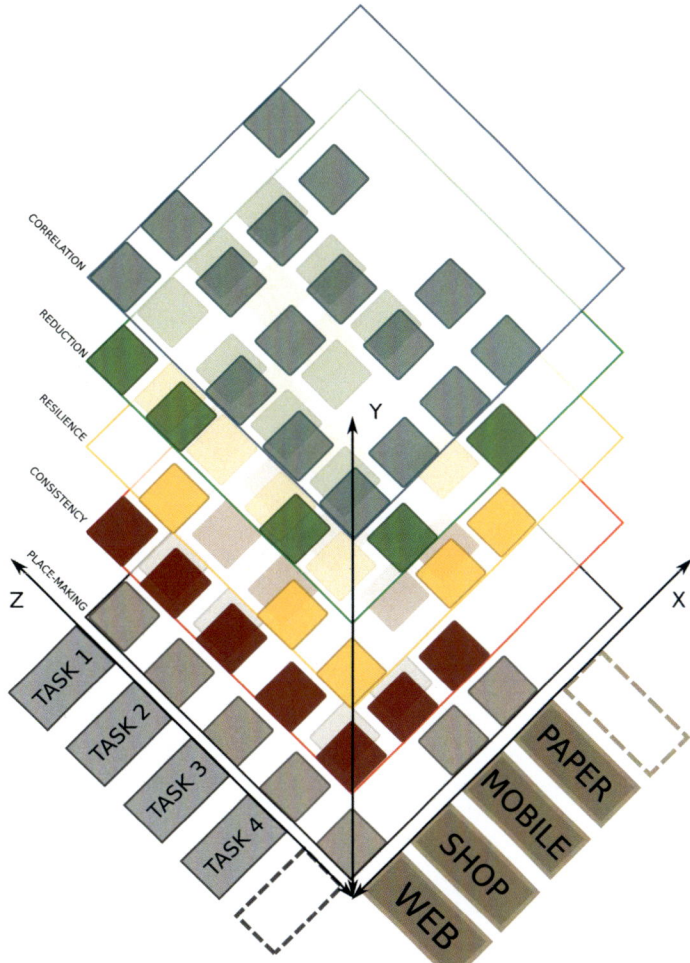

FIGURE 9.8
The CHU cube. The elements on the three axes are channels (C), here Paper, Mobile, Shop, Web, heuristics (H), and user tasks (U), here Task 1-n.

- that specific heuristic is not considered relevant in that channel
- that specific heuristic is not considered relevant for that user task: for example, *correlation* (as opposed to *place-making*) at log-in
- that specific user task is not relevant or even existing in that channel: for example (and for now), printing from a mobile application

These missing colored squares are to be investigated, as their being MIA could highlight some shortcomings that need to be addressed. Here an interesting visual characteristic of the CHU cube can really help: it allows slicing. Slicing and slices are useful ways to look for critical spots, missing or misdesigned user tasks, or local views over a specific problematic area.

If you think of heuristics, user tasks, and channels as lying on three axes in our three-dimensional space, you can see that you can get a rapid visual idea of the characteristics and pervasiveness of the information architecture by checking how the various layered colored squares (think of them as cubes) happen to be organized (Figure 9.9). Let's introduce a short-hand notation for the sake of simplicity: if *C* is for channels, *H* is for heuristics, and *U* for user tasks, we prefix those with 1 when they only touch on one element along any given axis and with *a* when they touch upon all elements of a given axis. *1U* then means that the slice is only relevant to *one user task*, and *aH* stands for *all heuristics*, meaning that this particular slice takes into account all five of them.

For example, the first slice to the left in Figure 9.9, *1U aH aC*, is what a cross-channel user task looks like. It's one user task but in all channels and with all heuristics accounted for. As such, we know that its characteristic of imprecision may translate in a number of "holes," squares/cubes that are empty,

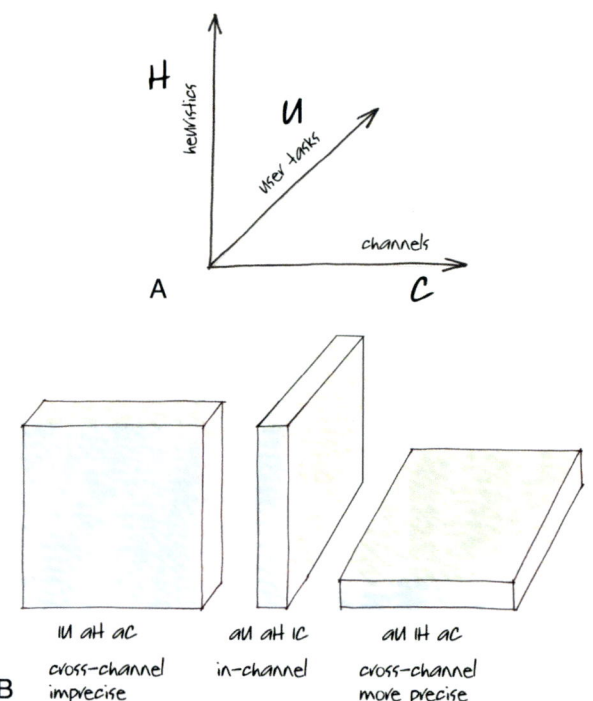

FIGURE 9.9

Slicing up the cube. It is interesting to note how the central slice is a basic rendition of Jesse James Garrett's diagram from Chapter 1 inside this model.

making the slice look like a piece of Swiss cheese, as some specific in-channel issues may not be addressed directly or may be addressed fuzzily to promote better global response to project goals. The horizontal, correlative axis is most prominent here.

It's still basically the same with the slice on the right, although this *aU 1H aC* slice simply represents how a single heuristic has been addressed for all tasks in all channels. Missing squares/cubes may signal local issues or the relative unimportance of a certain element. There is still imprecision here, although it is mitigated by the fact that we just have to deal with one single heuristic, which means that we do not have to accommodate for, say, improved *reduction* and plenty of *correlation* at the same time. As such, this slice is a useful control check, but does not represent any actual design strategy in our model. The horizontal axis is well developed here too, but being limited to one heuristic the slice leans toward precision.

Now take a look at the slice at the center: *aU aH 1C*. This is a one-channel slice: it represents how tasks and heuristics have been addressed on, say, mobile or the Web. As such, this is a general representation of the one slice that Jesse James Garrett described in detail in his *The Elements of User Experience*. It is what you would be looking into if you were in charge of the design of one single channel in a larger pervasive information architecture strategy. Taken in isolation, this slice is the one where design gets most precise: its horizontal axis is factually reduced to zero, and the heuristics work only internally. This is what we want to open up.

A few final notes. We usually build the CHU cube as a series of *Y* layers, by hand, on templates that basically work as checklists. We have each heuristic on a separate sheet and go through the various items checking or coloring them and taking notes when necessary (on the sheet). When we feel particularly creative, we use transparent sheets so that we can actually pile them up and build our 3D cube from 2D layers, very much like the cube in Flatland.

We work this way even when we are just designing parts of something that is not conceived as a cross-channel information architecture, but could become one over time. It is just to be expected that we will not be asked to design pervasive information architectures from scratch and from start to finish all of the time. Sometimes we will be in charge of a silo that we will try to pry open (think of Figure 9.9 and the *aU aH 1C* slice), sometimes we will be in charge of but a part of a pervasive ecology, and other times we will simply be adding new parts to an old, complex building that is being refitted to be pervasive. Whatever the case at hand, information is going everywhere and we design with that in mind.

SAMANTHA STARMER—DESIGNING INFORMATION FOR HOLISTIC EXPERIENCES

Until a few years ago, I never thought about designing for experiences beyond Web sites. I've been doing user experience and information architecture design for over 10 years, but my prior work for Amazon.com and Microsoft kept me focused on Web site experiences. My world opened up when I started at my current company and suddenly had to reconcile a customer's physical experience with the digital one.

REI (Recreational Equipment, Inc.) is a well-known and loved U.S. outdoor gear and apparel company. With over 100 brick-and-mortar stores, an adventure travel agency, outdoor events, related classes and advice, and active environmental stewardship efforts, we inherently provide experiences beyond just purchasing a product on a Web site. As a member-owned cooperative, we have a loyal customer base who loves nothing more than to talk climbing gear and share tips on where to find that sublime bike ride.

This sounds like a great situation for a company to be in, right? Strong fans, active community efforts, and a 70+-year-old physical presence that magnifies our digital opportunities. But managing such breadth of structured and unstructured information raises challenges. All of this information needs reconciliation and optimization across physical store signage, product tags, Web site content, marketing and advertising, catalog assets, mobile . . . the list goes on and on.

As technology advances and social media usage become ubiquitous, consumers are interacting with companies differently.

Digital and physical spaces are becoming blurred. We no longer have to drive to the bank; we can handle our finances from home or even on the train via our mobile devices. We hear about new products or promotions via our social network, and share our experiences via Twitter and Facebook. Consumers increasingly expect a seamless experience across all interaction touch points. Creating a seamless experience that optimizes the capabilities of print, face to face, Web sites, and mobile requires a new way of thinking and working.

This new way of working requires designing for pervasive information architecture. At REI, we need to provide consistent, findable, and discoverable information across all potential customer interactions. We must create cross-channel information architectures to support all of our content assets via any potential customer touch point, whether Web site, print, mobile, or face-to-face service interaction. Each touch point provides different capabilities and advantages, and relevant information should be optimized for each touch point.

Mobile phones have a small interface that is often used in quick bursts while en route to another location. Customers may bring product information printed from a Web site to use as reference in a physical store. New digital technologies allow for information to be updated continuously based on customers' location, behavior, and social networks. Information architecture for these interactions should provide a unified information scent across all channels and accommodate usage and environmental factors specific to the interaction. Information such as prices and product specifications should remain consistent, but an article about camping for beginners might consider content length and format appropriate to the device and environment the customer is interacting with.

At REI, moving from managing information in silos to more unified information architecture has required shifts in people, processes, and systems and in new ways of thinking about our technology, marketing, and experience visions. Accomplishing pervasive information architecture in support of a holistic customer experience is a journey. It does not happen quickly or easily; the direction sometimes seems to meander or even move backward. The ultimate destination will likely change as technology and consumer behaviors change. But there are active steps we are taking to adapt.

SAMANTHA STARMER—DESIGNING INFORMATION FOR HOLISTIC EXPERIENCES—CONT'D

Vision

REI needs to deliver personalized, relevant experiences that resonate regardless of how or where customers want to interact with us. This means aligning our vision across the company to support this kind of experience. In many companies, the physical and digital channels are trapped in organizational silos that rarely talk, let alone create unified visions. And yet, customers generally don't care about channels or departmental structure; they just want to enjoy their experience or complete their task in the manner of their choice.

In order to gain support for pervasive information architecture and the necessary people, process, and system changes, we have evangelized an overarching vision that focuses on the customer and their full experience across channels and touch points. This allows us to gain top-down buy-in and prioritization to facilitate conversations at all levels across the company. It also makes everyone accountable for improving the customer experience, which leads to broad support for pervasive information architecture.

People

We recently centralized our content creators under one team. This is a first step toward writing content that can be dynamically displayed and reused across channels. Previously, multiple groups were writing and rewriting information specific to a channel (in store, call center, Web site) or to a delivery mechanism (catalog, in-store signage, mobile device).

Recreation of information by unsynchronized teams can lead to poor customer experience if the information on the Web site is different from the information on a product tag or in the catalog. The majority of customers shop across channels and can become confused if the Web site says one thing and the product signage says another. Integrating the teams is a first step toward consistent information.

Processes

In moving to a "write once; publish many times" philosophy, we recognized that our information processes were dispersed through multiple divisions. We researched all of the relevant information processes and mapped out the gaps, overlap, or inefficient areas. Once we understood the pain points, we were able to recommend process optimizations that increase efficiency and reduce gaps.

We also created a core information architecture that focuses on how customers want to be able to find and discover information. We have used card sorting and other customer research tools to facilitate conversations with cross-divisional stakeholders about our Web site navigation, in-store way-finding, or product labeling. Noncustomer facing teams such as merchandising may have different needs for their information architectures, but we need to first focus on the best information experience for the customer and then link internal information architectures as appropriate to the customer-optimized ones.

Systems

These people and process changes would not be optimized unless we also reviewed the supporting systems. Starting with a unified vision across departments enabled us to understand needed system updates or replacements. Tools that provide sophisticated taxonomy and content management and that assist with information standards compliance can all be immensely helpful in providing efficient support for pervasive information architecture.

Through these efforts, REI is advancing from unconnected information within channels to true cross-channel experiences that are seamless across all touch points. Integrating our customer experiences and the supporting information architectures will allow us to dynamically present information that is relevant to our customers' needs and goals. Information is the foundation of our customers' experiences, and designing for all encountered information spaces requires that we develop holistic information architecture for the full information ecology.

Over the past 12 years, Samantha Starmer has worked on a wide variety of user experience and information architecture projects and strategy while at Amazon.com, SchemaLogic, and Microsoft. She is currently a senior manager at top U.S. retailer REI (Recreational Equipment, Inc.), where she is creating and leading new teams for user experience information architecture and interaction design. She is passionate about creating holistic, multichannel customer experiences and holds a Master's of Library and Information Science degree from the University of Washington where she regularly teaches on information architecture–related topics. You can find Samantha on Twitter as @samanthastarmer.

A CODESIGNED WRITABLE WORLD

In the language of today's computer geeks, we could call [our culture] a "Read/Write" ("RW") culture: . . . ordinary citizens "read" their culture by listening to it or by reading representations of it. . . . This reading, however, is not enough. Instead, they . . . add to the culture they read by creating and re-creating the culture around them. . . . As MIT professor Henry Jenkins puts it in his extraordinary book, *Convergence Culture*, "[T]he story of American arts in the 19th century might be told in terms of the mixing, matching, and merging of folk traditions taken from various indigenous and immigrant populations."

(Lessig 2008, p. 28).

We wrote in Chapter 6 that places, it does not matter whether physical or digital, are mnemonic palimpsests. They can be read as texts on which people sediment their stories and their interactions with other people, objects, and information. They can be written as well. Desire lines, pace layering, architextures, and information shadows—it's all about emphasizing an active role on the side of the users, who from passive recipients become active shapers, influencing the way the ecology evolves and performs. The five heuristics acknowledge this new role:

- *Place-making* establishes a clear difference between the concepts of space and place: the latter is a nongeometric entity that expresses experiential content, shaped and modeled by the interactions and emotional attachments of the people living it.
- *Consistency* introduces the idea of salience. Consistency is not an abstract measure, but a pragmatical and empirical goal resulting from the continuously changing pressure exerted by people and their needs, wants, and beliefs on the system.
- *Resilience*, *reduction*, and *correlation* all insist on the active role of the user, of the user's preferences, and of the user's behavior in shaping the final experience. Pervasive information architectures change, reshape themselves, and open up new paths in response to user action.

This is the general shift from a read-only (RO) culture to a read-write (RW) culture that Harvard law professor and free software activist Lawrence Lessig illustrates with the support of MIT's own Henry Jenkins, only we see it happen as a shift from RO to RW information spaces. Much of the real meaning that the label Web 2.0 purports reflects precisely that: a change in the way users participate in the process, a move from passive consumers to active coproducers. The social Web, collaborative and user-generated content, this is RW.

FIGURE 9.10

The urban performance "White Page" by Art Kitchen: a wonderful metaphor of a "writable world" and of user-generated information architecture. *Screenshot from Artkitchen.it.*

What we can see emerging though is a tad more radical, and probably moving beyond user-centered design into participatory design territory: a cocreated information architecture, *a crowd-sourced pervasive information architecture*, an entire ubiquitous ecology where designers and users share the responsibilities of creation (Figure 9.10).

This brings in legitimate concerns that in such a scenario the role of the designer might be diminished. And nobody likes design by committee. We don't see it as such. We see new strength, but fashioned differently and in a new dialogic context. There is less precise control, that's for sure, but a far wider imprecise opportunity to shape the use and reuse of vast ecologies of artifacts. It's all the difference between bidimensional Flatland and four-dimensional Spaceland, or Pervasiveland.

In pervasive information architectures, the design process becomes a holistic activity. It also moves from being just dynamic to being hybrid, as the Manifesto says (#4): pervasive information architectures embrace different domains (physical, digital, and hybrid), different types of entities (data, physical items, and people), and different channels. The boundaries separating producers and consumers grow thin, as do those among channels, media, and genres.

Information architectures also become a layered combination of top-down sketching and designing blended in with a perpetual bottom-up remodeling flow: the information architect lays down the fundamental bricks and connection rules, but users shape and reshape the building according to their needs, paths, and behaviors. It's again nothing more than the idea of users as wranglers Sterling (2005) talks about, with a little of Herman Hertzberger and his Diagoon housing (Chapter 3) thrown in for good measure.

In a way, we could say that pervasive information architectures cannot be designed from top to bottom by a single talented individual or by a dedicated small group of professionals. They cannot be designed in one go, as well. They are complex, iterated systems, and the role of the designer is mainly that of the enabler. The information architect provides the rules of the game, the board, and a little coffee. The users play the game in its infinite variations, building their strategies, their paths, and their experience. And enjoy the coffee.

If you think this is limiting on the designers, consider your own house or apartment and see where the influence of the architect ceases. Does your sitting room have the same furniture and the same paint on the wall that all other apartments in the building or houses in the neighborhood have? We bet it does not. And how many personal, intimate, cherished but probably useless and conflicting items have you brought in through the years? Let's see. Is your apartment in central Paris? Then what does that African statuette do in your hall? Oh, you traveled to Ghana and saw that somewhere so you bought it and took it home. Good. Is your house in the greater Chicago area? And you enjoy your Japanese bedroom and Shaker kitchen? Oh, we see, you really love sushi but appreciate Americana.

As you see, thinking in terms of Stewart Brand's pace layering (Chapter 6) is the key here. And it does not seem that architects ever felt dispossessed because they could not decide how you were to choose your table and chairs.[4] See it this way: if in Web 2.0 times it is user-generated content and folksonomies, in ubiquitous ecologies it's going to be user-generated architectures in ubiquitous ecologies. Ubiquitous ecologies are complex open systems: this implies that the way they evolve is not entirely predictable. We start them up, we do maintenance, we tweak the engine, but that's it. In this continuous metamorphosis and perpetual precarious balance between shaping and reshaping, writing and rewriting, pervasive information architectures exchange control for richness, mainstream for participation, certainty for innovation.

So should we, their designers.

---

[4] Which is not to say that many of them, and many of the big names, actually tried to do precisely that.

KARS ALFRINK—THE CITY IS MY GAME CONSOLE

campaign for their movement at these places. They were scored based on the number of followers they got. And the winner got cash and coaching to make their movement a reality.

The game was designed to have them experience the value of collaboration first hand. It was also used as a visual indicator of what was going on in the city during that year. And it transformed an area of Rotterdam, which is usually almost exclusively used for shopping, into a political arena, sucking in pedestrians and redefining the relationship between young people and adults.

Some of the stuff I find most exciting to work on at the moment is hyperlocal game design.

> In the end, the design of technology . . . must let us actively practice at something, however humble. Taking part in locale is one such activity.

I run a studio that designs games for public space. To give you an example, last year we did a game for the European youth year, which took place in Rotterdam.

How it worked was these kids were all starting movements. They competed for territory by planting flags. They could then

That departs from the aforementioned quote from Malcolm McCullough's book *Digital Ground*, where he argues that technology, urban computing if you will, should facilitate people's participation in place-making.

FIGURE 9.11
Reflections. Photo:
M. Glullano D. M.
Source: Flickr.

Continued

KARS ALFRINK—THE CITY IS MY GAME CONSOLE—CONT'D

Because, to some extent, many urban spaces have become just that, space, without any history, layering, or localness to them. They could be anywhere. And so McCullough argues for designers to be sensitive to place and deploy technology in a way that is appropriate to it.

And I think that to a large extent what has been going on with games in cities, often at least, is that they don't really relate to the specifics of the location and that is a shame. Because games can be tools to "re-place space," if you will.

So maybe, another example of a game we did will help clarify the point I'm trying to make. It was called Koppelkiek. We ran it in a troubled area of my hometown Utrecht, called Hoograven. It was commissioned by a design event that looked at the function design can have for society.

We were inspired by Jane Jacobs's ideas about the charms of city life being the many interactions with strangers, which runs against much of the contemporary thinking about integration issues, which basically says we should all become best friends forever.

We came up with a game that would gently encourage casual interactions in the neighborhood through a very light-weight rule set that would run pervasively over a period of three weeks. The basic idea was: you take photos of yourself with others in various situations for points. We came up with assignments that were place specific.

What was interesting was that people were relieved about something happening in their neighborhood that wasn't about the problems there. But instead it was just something different from the stuff that was usually going on (which wasn't much).

I think of this game as a way to kind of amp up the diversity of uses of the streets—again inspired by Jane Jacobs and her thoughts about the emergent, complex order of city life. I think there's a real role for urban games there. And it is one that is at the core of why I started Hubbub. I don't want to see streets be used just for shopping and commuting. There's more to life than just this.

I also enjoy thinking about how you can use games to achieve local effects, but not by forcing them onto people by submitting them to arbitrary rules and telling them it's a game. That's bad design. There can be a loose coupling between a game and its second order effects, as discussed before with *Change Your World*, which is mostly about skills and attitudes.

But another aspect of a game such as Change Your World, and many other event-based games, is an effect similar to what is common practice in the world of culture jamming: the temporary autonomous zone. Carnivals and block parties all to some extent fit in this category.

FIGURE 9.12
Photo: G. Emel.
Source: Flickr.

KARS ALFRINK—THE CITY IS MY GAME CONSOLE—CONT'D

This effect comes about through a mutual agreement on rules. Actions and interactions in the city get new meanings. When speaking of this dynamic, game designers use the term *magic circle*.

Take boxing for example. Within the artificial reality of the boxing ring, punching someone in the face gets you points. Doing the same outside of the magic circle of boxing, on the street, would likely get you jailed.

There's this really interesting dynamic between passersby not in the know and game players. It is part of the fun. Geocachers, for instance, really enjoy doing something out of the ordinary in the city that no one notices. This reminds me of the hobo code. It is a very effective pattern.

Other players, such as those engaged in a game of capture the flag, enjoy the fact that people are startled by their odd behavior. This is play as performance and when done artfully can make a game in a spectacle that is as enjoyable to watch as it is to play.

There are many aspects to game design. When you add place specificity to the mix it becomes even more challenging than it already is. You need to immerse yourself in the environment. We set up a temporary studio in a vacant shop when we were doing Koppelkiek. We sought out community leaders to have them be ambassadors for our game. We ran play test in situ and so on.

We also walked the area many times to get a sense of the systems and the processes that were already there. We did this before we even started designing the game. We've been very inspired by the vocabulary developed by Kevin Lynch for this. It provides you with a much more fine-grained view of how a city is experienced.

To bring this back to architecture and city planning, what I find very exciting is that urban games pose real challenges to those disciplines. They demand them to plan for the unexpected to allow for enough space and looseness for play to happen.

This is commonly known as *adaptive design*, allowing people to reappropriate their devices, environments, etc. And you know what? Game designers could really offer some help there, because like cities, their rule sets need to allow for play, they can't be too tight (no choice, you're railroaded), or too loose (confusion, no meaningful choices). So I guess, ultimately, the effect I hope we can achieve with these kinds of games is enhancing the autonomy of the urbanite.

Kars is an independent interaction and game designer. His main professional interests are cities, physical and social interactions, and play. He is also a teacher and an organizer of events. He lives and works in Utrecht, The Netherlands.

FIGURE 9.13
Photo: J. Lucas.
Source: Flickr.

CASE STUDIES

The Pervasive Supermarket

After so many examples dealing with eating, drinking, or buying food, it seems only natural that we try to recompose the picture by seeing how we can apply everything we have been discussing to the design of a pervasive information architecture serving a supermarket. Call it poetic justice, but a supermarket makes for a very interesting example for a number of factors: it is something we can all relate to, as we all know how it is to go shopping for groceries; it is a place where we spend a fair amount of time; it usually deploys multichannel strategies, in the store, on paper, or on the Web; and, traditionally, it is the playfield for expertise other than that of information architects.

Let's start from this last point: oddly enough, given the sheer amount of information they process and send to the customer, the very idea of an information architecture is often either undeveloped or totally absent in the design of the physical retail store. This is a point to be addressed. Browsing the rather large bibliography on the topic is an exercise in a very long list of individual points of view: a single global vision of the shop as an encompassing experience is nowhere to be found. Professionals usually involved in the design cycle (architect, interior designer, marketing, and advertising) do not operate as a team or share a common vision of the user experience, with a net loss in complexity and global customer satisfaction.

This way, easy solutions or traditional solutions become the solution. The design of the brick-and-mortar store is reduced to matters of simple choices—let us have a large, rectangular open space to facilitate movement—basic logistics—how many shelves, what shape of shelves, what kind of aisles—and marketing—how to position the products on the shelves. Then some attention is devoted to branding.

Do not misinterpret us. These are important themes and proven approaches, but if dealt with in isolation they steal the scene, impose a frame, blur the big picture, and make real innovation more difficult. The supermarket today goes way beyond the aisle and shelf or the brick-and-mortar store: it involves the complex network that ties customers to the chain and the brand in every channel they are active in.

User Experience as a Process

User experience is a process: it is a dynamic phenomenon that evolves in space and time. In the design of a cross-channel, pervasive information architecture for a supermarket, the first necessary step is to acknowledge that shopping there will never entirely coincide with just the single act of buying products and viceversa. Shopping is not just picking up something from the store or the Web site and paying for it. Different places and contexts play a part: reading a weekly flyer of special offers, being sent an e-mail about some upcoming sale, or hearing from friends or on the radio.

To make this example more manageable in the little space a book offers, we will limit our case study to four sample channels (C) we identify as:

- paper, meaning all communication done via printed materials, including flyers, brochures, and ads
- store, meaning everything that has to do with the physical store, including shop layout, signage, way-finding, shelving, and positioning, among others
- Web, meaning all communication passing through the Internet, with the exception of that which becomes mobile, including Web browsing, mail, and the download of software updates, for example
- mobile, meaning all communication happening on mobile devices via specific applications or systems.

Because we are considering a supermarket as our scenario, we might resort again to our dear Mrs. Hutchinson from Chapter 1, as she is sure in need of some food for the house after that weekend in Florence. All major chains send advertising in the mail on Fridays and Saturdays, so she has a good deal to read through. There are a few products on sale this week and a couple of promotions and special offers she might be interested in. She also needs some of the basic stuff, as the fridge is almost empty after a week-end with the family on the prowl without her. She compiles a list and logs in to the supermarket Web site to see if she is eligible for any discounts this week. She is, and she prints out a couple of coupons. Then, during her lunch break at the office, Mrs. Hutchinson drives to the store. She finds a parking place on a remote corner of the lot: everyone seems to be here today. She fetches a cart, enters the store, and starts to shop.

This is an almost automatic process, as she knows the place and knows where the stuff is. But she can't find a couple of the offers and has to ask the staff. When she is done, she queues at the cashier, pays, and is out. Her break is almost gone: she hurriedly moves all her bags to the car and she is off.

There are a number of steps involved here, all of them important to the final user experience, all of them with relevant information-based aspects attached. In bullet points:

1. Getting information about current products and offers
2. Building a shopping list
3. Getting to the supermarket

4. Parking the car[5]
5. Fetching a cart
6. Navigating the shop
7. Finding products
8. Queuing and paying

These steps are our sample macro user tasks (U). If this was a traditional super-market, all tasks except possibly the first one would be tied to one single chan-nel, that of the store and its physical layout. But we want to break the silos, and this supermarket is pervasive. So Mrs. Hutchinson has been reading the flyer (C: paper); is a registered user on the supermarket's Web site (C: Web), which provides her with special discounts; and has a mobile application that she uses for finding products and paying (C: mobile). And of course she needs to find a parking place and navigate the shop (C: store). We also want to consider the fact that (1) going home and (2) unpacking and using the products are actually part of the shopping experience (Mrs. Hutchinson is not going to be happy about our supermarket if the meat is stuffy or tasteless or if the yoghurt is watery).

All of these cross-channel activities influence and ultimately constitute the cus-tomer experience of shopping in the supermarket, not as the simple mathe-matical sum of all of the micro- or macroexperiences characterizing each task, but as an open, complex, and dynamic system. So let's see how each heuristic can help us create seamless pervasive information architecture across the four different sample channels for a more satisfying user experience.

Place-making

Building a sense of place in our supermarket is a cross-channel, high-level activity. It means conceiving pervasive information architecture in the first place, a con-sistent and stable information architecture model that works across the differ-ent channels of this specific ecosystem. It means supporting information scent and berry picking across the channels, for example, by providing custom navi-gation that follows the user, preserving tone, language, structure, and appear-ance and suggesting thematic paths in the flyer or on the Web site that are then easily recognizable in the store (Figure 9.14).

Place-making should be used for building a pervasive way-finding system so as to guarantee a seamless experience even when transitioning from one channel to another and in shaping the various channels. This is again one specific aspect of the more generic dialogic nature of external place-making versus internal

---

[5] Mrs. Hutchinson could surely use some public means of transport to get there. This would change the specific user tasks from parking to knowing which is the right stop for her, for example. This being a sample scenario, we wanted to make it as simple as possible. So Mrs. Hutchinson is one hell of a driver.

FIGURE 9.14
Enhanced cross-channel way-finding reinforces a sense of being there and helps customers feel at home. Image: A. Falcinelli.

place-making: while the former builds a more imprecise sense of place across all channels so that you can recognize the supermarket and discriminate the bakery from the fresh produce department both on paper and then in the store at a glance, the latter works to provide the best environment channel per channel.

Once the system supports customers through pervasive information architecture, there is no need to stick to a run-of-the-mill physical layout in the store either: this is now an augmented place, connected seamlessly to the other channels:

> [its] general layout might even move away from the common linear, sequential layout where a number of straight racks are placed back to back, for example to a radial structure where a core, a main hall, acts as the entry and exit point and is used as a multi-purpose landmark. It is the main physical way-finding hub, so it is easily accessible from every entrance and from every area. This is home, and it acts as a one-stop shop which allows fast navigation toward all departments, racks and shelves, cashiers, check-out points: It accommodates all starting points for signage, those marking personalized navigation and those for theme paths, discount sales, maps, interactive indexes, accessible displays, and help desks . . . . In this supermarket aisles become lighter artifacts. Showcasing and physically acquiring products, especially for certain kinds of goods, might become two separate activities. The shelves themselves might just as easily be simple displays, with no

> more than a few items available for direct evaluation: samples, whose code customers can read into their devices to acquire all necessary information. Later, after they have been confirmed, the real items could be checked out from a larger warehouse area closer to the parking lots.
>
> (Resmini & Rosati 2010, p. 95).

All channels provide a part of the general hodological identity of the supermarket, reinforcing the sense of place and helping users feel at home. Visual and cognitive cues are bounced from channel to channel, preserving their structural and logic continuity and varying, when necessary, their appearance, as we saw Jean Jacques Annaud do in moving the library in *The Name of the Rose* from page to screen in Chapter 4.

Consistency

The pervasive information architecture of the supermarket has a different use for internal consistency, that is, the capability to provide a coherent in-channel experience that suits the context, goals, and users it is designed for, and external consistency or the capability to support and sustain the same logic across all channels.

Designing internal consistency means adopting and maintaining pragmatically correct classification systems—not theoretically sound ones: this is classification for the staff and customers who have to use it. As such, it should not, for example, be company centered or adopted from external sources as is. It also means that the organizational schemes in place should support different approaches to products and information, different styles of searching, and different shopping behaviors. This is deeply connected to *resilience* as well.

Designing external consistency means compromising where the architecture needs to provide a sufficiently globally accurate view for graceful transitions from one touch point to another and from one channel to the other. Being precise here very often means losing continuity across channels.

In the early stages of design we often adopt a generic scheme that works as a mixed classification model (a scheme within the scheme) and that we tweak in subsequent iterations. This is based on

- a hierarchical-enumerative model on the first level (a taxonomy)[6]
- a faceted model on the second level

---

[6] It might be worth noting that for niche or specialized retail stores, those selling a more homogeneous range of products, it might be enough to use one single general faceted scheme (a prerequisite for the applicability of such schemes being that all items to be classified share similar characteristics).

The faceted scheme on the second level allows us to address several mental models at the same time, enabling users to access the elements of a given collection from different points of view, in this case, the products and related pieces of information in our supermarket.

Faceted classification is widely used today in applications such as iTunes, on devices such as the iPod, and on Web sites, where it seems to be all the rage for online shops. On the Web it is often simplified and used as part of a general, at times implicit, taxonomy. A good example of a classification system mixing taxonomies and facets on the Web is the grocery market at Amazon Grocery & Gourmet Food (Figure 9.15).

Because quite a few patterns are available (Hearst 2009; Kalbach 2007; Morville & Callender 2010), it is not difficult to imagine ways to apply such a scheme to the logic of the store. Actually, even if slightly unusual, it's not hard to imagine

FIGURE 9.15

Amazon Grocery & Gourmet Food employs a mixed classification system consisting of a full-fledged taxonomy at the first level and a faceted classification at the second level for improved findability.

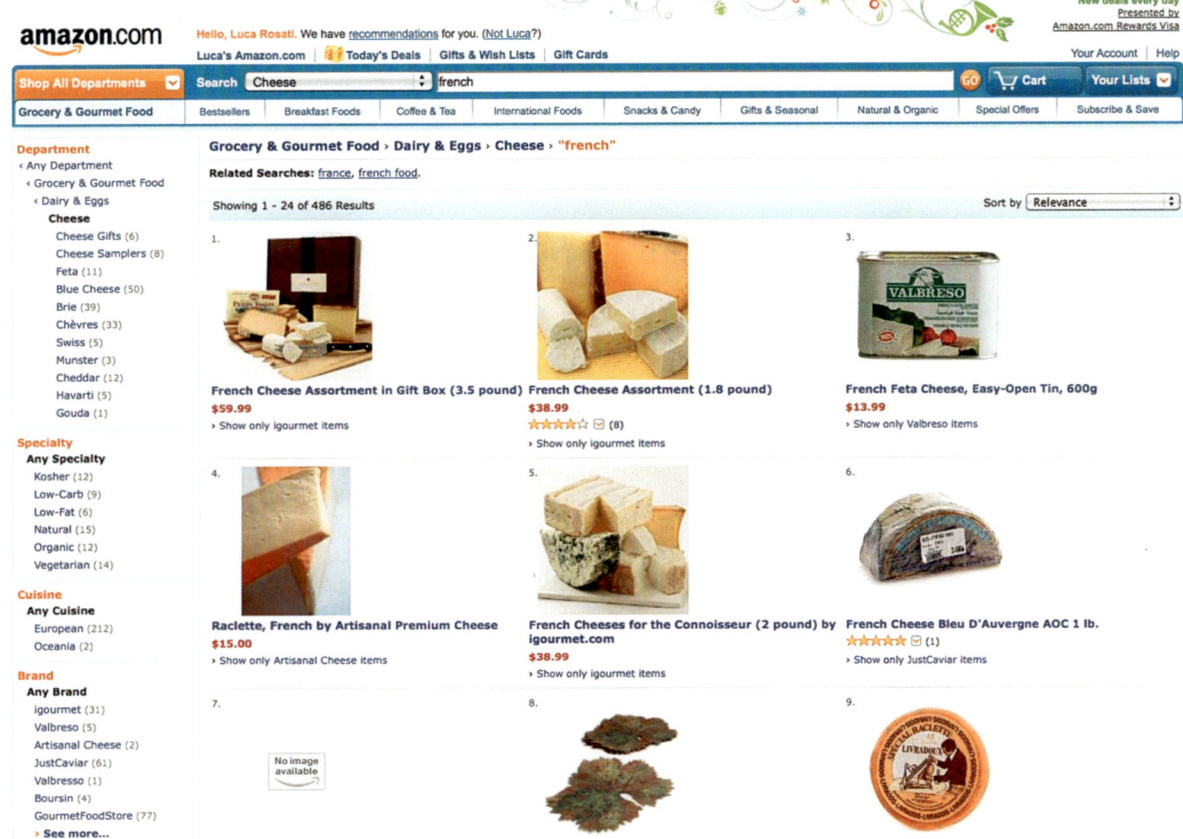

some degree of faceted classification adopted in paper as well. After all, facets come from the world of paper and books, where they were invented to solve categorization issues in the analog world of libraries.

From paper to the store, chromatic, iconic, or alphanumeric codes may be used easily to identify the facets and to favor matching mechanisms, recognition, and recall. Mobile devices tied into the pervasive layer of the supermarket could reuse these codes and reinforce consistency, bringing the hypertextual logic proper of the Web into the physical environment (Figure 9.16). The easy manipulation of informational facets could introduce the possibility of experimenting with product placement and layout, for example, by having facets for the different meals in the day, regional recipes, or food lifestyles.

Resilience

We said that in pervasive information architectures the stories of interactions between users and information, users and places, and users and other users should be used to shape or reshape the places themselves in real time. Users are codesigners, and places are palimpsests, rewritable memory receptacles that record paths, choices, and interactions.

This is what we already see on the Web: push strategies, analytics, task tracking, and social and collaborative tagging have long accustomed us to an environment that learns from our choices, preferences, and strategies and tries to anticipate them.

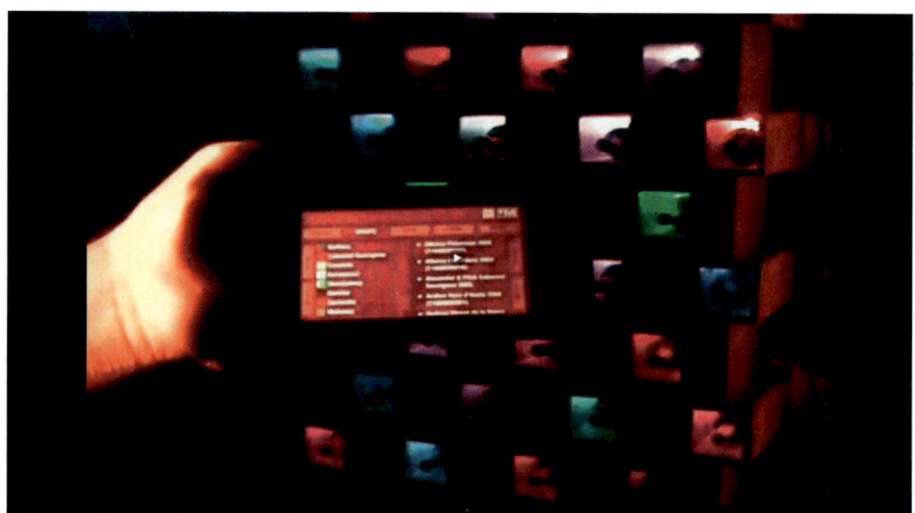

FIGURE 9.16

WineM by ThingM: an example of faceted classification applied to retail. *Screenshot from Thingm.com.*

Think about user tasks 6 and 7, navigating the supermarket and finding products. These are user tasks where the primary channel is the store. How much of these Web-only strategies are actually applicable there? How much of Amazon can we transfer to the store?

Some of it could work as is, with minimal technology, and it already does, as explained in Chapter 8: receiving meaningful push suggestions and being provided with interesting correlations are something stores have been doing forever. But in the pervasive supermarket, more could be supplemented easily via the mobile channel: imagine your phone being able to use some sort of supermarket app (Figure 9.17) that helps us

- refind and retrace paths and shopping tasks we have already used in the past, regardless of the channel in which they were created
- customize our shopping experience by checking our Web profile, if any, or our past shopping carts, allowing us to save time and money
- share our histories and profiles with family, friends, or peers

FIGURE 9.17

Use of mobile computing in the pervasive supermarket helps refind paths, customize the shopping experience, and receive ad hoc suggestions. Image: A. Falcinelli.

In our pervasive perspective, however, there is more: we have to ensure that a dialog is there between channels in terms of user tasks, as behaviors registered on the Web should be retrievable in the store, or vice versa, and that user tasks are used effectively to increase the capability of the system to answer emergent needs. One way to achieve this has been described in Chapter 6, when introducing the *resilient museum*, and involves the use of a service middle layer that acts as a sieve, filtering the fast moving customer-provided information and data into the slower top-down structure of the supermarket's information architecture, increasing the complexity of the system.

Reduction

We characterized *reduction* as the capability of an information space to minimize the cognitive load and frustration associated with choosing from an ever-growing set of information sources, services, and goods. We said this is not the same thing as discarding options to reduce the number of choices available, but rather organizing and presenting choices in a way suitable to the context and the users.

In large retail contexts, such as that of our pervasive supermarket, the paradox of choice can be counteracted by segmenting all available options, employing both the *organize and cluster* and the *focus and magnify* strategies discussed in Chapter 7.

Again, adopting a mixed classification system is a good way to go.

The taxonomy at first level serves the main departments of the store, whereas the faceted classification at the second level segments products belonging to the same department. While the former allows customers to find their way and address their choices to a specific department, the latter, however, allows satisfying multiple seeking strategies and needs at the shelf level. This is *organize and cluster* in the store.

Facets are not only useful when accessing information for the first time, though, but also when tweaking the seeking process. Facets are a wonderful tool for refining a search, suggesting related products, or eliciting latent needs; on the one hand, the combination of a taxonomy with facets enables customers to focus their attention early on and reduce the amount of unnecessary choices; on the other hand, facets allow people to extend their shopping to related products either by similarity, by coupling, or by using social patterns. This is *focus and magnify* (Figures 9.18 and 9.19).

FIGURE 9.18
A focus and magnify strategy (step 1) reduces the paradox of choice, allowing customers to focus their attention only on those products that match their interests or profiles. Image: A. Falcinelli.

FIGURE 9.19
A focus and magnify strategy (step 2) reduces cognitive overload by allowing people to expand their choices only after having first zoomed in to their niche of interest. Image: A. Falcinelli.

Correlation

Correlation is the capability of pervasive information architecture to suggest relevant connections among pieces of information, services, and goods to help users achieve explicit goals or stimulate latent needs. Internal correlation links artifacts belonging to the same channel; external correlation links artifacts across channels.

Correlations are both top-down, built into the information architecture by the designers or the administrators, and bottom-up, added, either actively or passively, by users and customers.

Correlation is mostly a digital affair as of today. The store is still largely a vertical place, where the hierarchical and static placement of products in terms of departments, aisles, and shelves is vastly prevalent on the horizontal dimension that correlates artifacts syntagmatically.

> Usually, product layout in a shop is firmly tied to racks, shelves and aisles and is market-driven, with little attention devoted to a global vision in terms of user needs and user navigation: the very idea of way-finding (Lynch 1960) which is strategic for a successful mapping between the physical environment and our cognitive perception of it (Resmini 2007) seems to be somewhat relegated to large architectural spaces such as airports, stations, malls, and usually only for macro-navigation and for taking us from one place to another. . . . RFID, touchscreen technology, mobile phones, . . . are actively increasing the amount of data we (or our enhanced selves) produce, receive, process and transmit: these data could be used to improve the relational dimension of information architecture in physical spaces, allowing for related-items links among products or families of products and alternative ways of navigation which free themselves of the shelf structure such as theme paths, coupling paths (i.e., "if you bought a we suggest b and c, that you may find there"), and recommended or best selling products.
>
> (Resmini & Rosati 2008, p. 10).

For example, one possible way out of the trap is by breaking the silo again and offering correlations between products on the mobile channel. These could tap into the reviews or suggestions users are providing on both the supermarket's own Web site and the general Web, using the most diverse sources to provide:

- special paths, for example, vegan food or ethnic cuisine
- coupling paths, such as "if you like a, we suggest b and c that you may find there"

- correlations through faceted classification, highlighting items belonging to the same facets
- custom correlations based on customer profiles or previous shopping experiences
- social correlations based on social-collaborative patterns
- cross-channel correlations targeted at refinding products, such as connecting an offer on the Web to the actual item in the store

Going CHU

All of these midlevel, information-based design considerations can be translated into very specific CHU (channel–heuristics–user tasks) issues. This is actually how we usually proceed, by mapping these indications to more precise actionable items.

Considering that the whole superset of user tasks outlined on pages 215–216 would require an entire new chapter just to get started, we think we can illustrate the process safely and successfully with the help of one single macro task across all heuristics and channels and have you survive the ordeal. This way, we are considering just a slice of our CHUbe, and precisely the left-most slice in Figure 9.9, the *1U aH aC* slice. If you remember the short list back at the beginning of this chapter, our sample channels are the store, paper, the Web, and mobile. And let us agree that our user task for this example will be navigating the shop.

In many projects, but not all, of course, one of the channels will be prioritized because it is the most developed, the most used, or the most difficult to redesign. For the sake of a silly example, think about waking up one day with the idea that you will base your redesign of all cross-channel communication for the Coca-Cola Company to be based on a green palette. Not a wise move, is it? Coca-cola is red, and hence a number of subsequent design choices will be flowing downstream from here: you will have one channel that you consider somewhat your master mold, or your primary reference point.

In our example, this channel is the store. We can redesign the remaining channels at will.

This means, for example, that paper, the Web, and the mobile channel will largely first inherit, adapt, and modify a number of base proprieties from there and then feed any necessary changes back to the store itself. Different channels can be different master molds for different characteristics, of course.

This is a slice, so you won't mind if we dribble all visual concerns we have with the CHUbe and simply render it as a table. Here we go.

Table 9.1 User Task "Navigating the Shop" Rendered in Table Format for Our Sample *1U aH aC* slice (Heuristics on the *Y* Axis, Channels on the *X* Axis)

	Store	Print	Web	Mobile
Place-making	Color code aisles and shelves. Provide visible unambiguous visual cues. Consider different types of users and mobility.	Respect color codes. Respect the physical layout of the store. Tie individual products to both their placement in the store and their findability online. Provide maps.	Make people feel at home. Translate the physical layout into clear information architecture. Use established labels, colors, and fonts. Provide maps.	If providing in-store navigation, rely on established conventions for labels, colors, and fonts. Augment the sense of place with in-context information when in-store. Provide premade paths.
Consistency	Use colors, surfaces, labeling, and signage in a way that considers shopping and user goals. Verify labels and names and use them consistently.			
Resilience	Allow more than one path to products. Do not build one-way paths. Allow for shortcuts. Consider variations from the standard aisle model.		Allow people to customize their experience, but do not vary the mental model established with place-making and in other channels. Store preferences. Offer more than one way to find information.	Provide people with histories of their choices on the fly. Make these aware of current choices and offer suggestions for missing/needed products.
Reduction	Use the physical layout of the shop and signage to build meaningful clusters of products. Use aisle height.	Only print those products and lines of products that are relevant for the current customer/time/supermarket. Link to other channels for more.	Provide the full catalog, but use strategies to offer information incontext.	Be contextual. Mobile moves with the customer. No need to provide info on products not currently available, for example.
Correlation	Place products where they are needed, even if this means having them around.	Suggest theme, goal, and price couplings.	Suggest best buys, couplings, and social recommendations and allow printouts.	Link products according to social and on-the-fly input, for example, from staff (sales or sold-outs).

Did you read it all? Good. It's certainly not exhaustive, but then you can see we were serious when we said that complexity really has a way to sneak in and is largely unavoidable. All of these operative instructions turn into more and more specific design requirements as you move in. Depending on the project and your degree of involvement, you might turn these into full-fledged specifics for XHTML or the building of a new signage system across all channels, or stay on the surface and release generic redesign guidelines. Whatever you do, this is simply the starting point, the first step out of the door. Walk on: it will be an interesting journey.

More Tables

The following tables provide some more examples of possible strategies and related design tools and solutions that address the middle ground between the overarching cross-channel pervasive information architecture model and the CHU systematization. Again, this is by no means an exhaustive list and is only provided for reference and further reelaboration.

Table 9.2 Place-making across Channels

Channels	Possible Strategies	Sample Design Tools
All	Common information architecture model and identity. Ease to share/find/refind items across channels	Analogical: maps, colors, icons, letters, and numbers. Digital: QR codes, RFID, or other unique digital identifiers + mobile or fixed digital devices
Store	Berry picking and information scent	Analogical: maps, colors, icons, letters, and numbers. Digital: QR codes, RFID, or other unique digital identifiers + mobile or fixed digital devices
Print	Berry picking and information scent	Colors, icons, and so on to suggest thematic paths (e.g., special pasta; Piedmont wines), correlations between products (couplings, similar items)
Web	Berry picking and information scent	Thematic paths, contextual navigation; social and custom navigation
Mobile	Berry picking and information scent	Thematic paths, contextual navigation; social navigation; personal navigation; in-place cues and directional aids

Table 9.3 Consistency across Channels

Channels	Possible Strategies	Sample Design Tools
Store	Adoption of a mixed classification system or a simple faceted classification system	Analogical: icons, colors, alphanumeric codes to identify departments, aisles, shelves Digital: QR code, RFID or other digital codes + mobile devices
Print	A selection of categories (taxonomy) and facets used on the Web and in the store	Colors and/or names to identify main categories of the taxonomy; colors, icons, and symbols for main facets; eventually also RFID and QR code
Web	Adoption of a mixed classification system or a simple faceted classification system	Patterns for faceted classification on the Web
Mobile	Adoption of a mixed classification system or a simple faceted classification system	Patterns for faceted classification on the Web

Table 9.4 Resilience across Channels

Channels	Strategies	Tools
Store	Searching	Digital way: mobile or in-store devices allow locating departments or products easily by electronic codes (RFID or similar). Analogical way: iconic and alphanumeric codes make the same
	Browsing	Departments, aisles, shelves clearly findable using digital IDs or analogical coordinates
	Monitoring	Push suggestions according to the user profile (via mobile or in-store devices) Custom paths for returning users or specific targets/ needs Wish lists
	Being aware	Popular items or paths; related items
Print	Searching	A–Z index. Highlighted items
	Browsing	Grouping items according to their location in the store (departments, aisles) using iconic or alphanumeric codes
	Monitoring	Receiving only print advertising that matches specific profiles or needs
	Being aware	Popular products. Offerings and sales
Web	Searching	Search engine, A–Z index
	Browsing	Main and local navigation. What's new
	Monitoring	RSS, newsletters, wish lists
	Being aware	Social navigation; custom navigation (history, profiled suggestions, etc.); contextual navigation (related items)

Table 9.4 Resilience across Channels—Cont'd

Channels	Strategies	Tools
Mobile	Searching	A–Z index, search functions
	Browsing	What's new. Local, contextual navigation. Main navigation
	Monitoring	Wish lists. Suggestions from social contacts
	Being aware	Social navigation; custom navigation (history, push suggestions etc.); contextual navigation (related items)

Table 9.5 Reduction across Channels

Channels	Possible Strategies	Sample Design Tools
All	Organize and cluster; focus and magnify	Adoption of a mixed classification system

Table 9.6 Correlation across Channels

Channels	Possible Strategies	Sample Design Tools
All	Opportunity to share/find/refind items across channels	History or wish-list QR codes, RFID, or other kinds of digital codes coupled with mobile or fixed digital reading devices
Store	Apply information-aware strategies	Maps, colors, icons, letters, and numbers for navigation. These can be both physical and digital
Print	Faceted classification or social organization provides alternatives to hierarchies	Analogical: as above Digital: QR codes (RFID)
Web	Faceted classification (top-down) coupled with social classification (bottom-up)	Patterns for faceted and social navigation. Theme and custom paths, contextual navigation; social navigation and collaborative tagging
Mobile	Faceted classification (top-down) coupled with social classification (bottom-up)	Patterns for faceted and social navigation. Theme and custom paths, contextual navigation; social navigation and collaborative tagging

RESOURCES

Articles

Institute for the Future. (2000). The Future of Retail: Revitalizing Bricks-and-Mortar Stores *Institute for the Future Corporate Associates Program*, *11*(1).

Kopalle, P. K. (2010). Modeling Retail Phenomena. *Journal of Retailing*, 86(2), 117–124.

Kourouthanassis, P. E., Giaglis, G. M., & Vrechopoulos, A. P. (2010). Enhancing User Experience Through Pervasive Information Systems: The Case of Pervasive Retailing. *International Journal of Information Management*, 27, 319–335.

Layout. (2009–2010). *Mark-Up*. http://www.mark-up.it/articoli/0,1254,41_ART_4506,00.html.

Morville, P. (2010). Ubiquitous Service Design. *Semantic Studios*, April 19. http://semanticstudios .com/publications/semantics/000633.php.

Potente, D., & Salvini, E. (2009). Apple, IKEA and Their Integrated Architecture. *Bulletin of the American Society for Information Science and Technology*, 35(4), 32–42, April/May. http://www .asis.org/Bulletin/Apr-09/AprMay09_Potente-Salvini.pdf.

Resmini, A., & Rosati, L. (2010). The Semantic Environment: Heuristics for a Cross-Context Human-Information Interaction Model. In Dubois et al. (Eds.), *The Engineering of Mixed Reality Systems* (pp. 79–99). Springer.

Wikipedia. (2010). Syntagmatic and paradigmatic relations. In Wikipedia, *Course in General Linguistics*. http://en.wikipedia.org/wiki/Course_in_General_Linguistics.

Books

East, R., Wright, M., & Vanhuele, M. (2008). *Consumer Behaviour: Applications in Marketing* (specifically Chapter 10, the Retail Context). Sage.

Friedberg, A. (2006). *The Virtual Window: From Alberti to Microsoft*. MIT Press.

Iacobelli, G. (Ed.). (2010). *Fashion Branding 3.0: La multicanalità come approccio strategico per il marketing della moda.* (Multi-channel as a strategic approach in fashion marketing). FrancoAngeli.

Lessig, L. (2008). *Remix: Making Art and Commerce Thrive in the Hybrid Economy*. Bloomsbury.

Levy, M., & Weitz, B. (2008). *Retailing Management*. McGraw-Hill/Irwin.

Underhill, P. (2008). *Why We Buy: The Science of Shopping*. Simon & Schuster.

Movies

Metro Group. *Future Store Initiative*. http://www.future-store.org; http://www.youtube.com/ watch?v=j0k5_CQPx_U.

Microsoft Office Labs. *Retail Future Vision*. http://www.officelabs.com/projects/retailfuturevision/.

References

All digital resources last accessed December 20 2010.

Abbott, E. A. (1995). *Flatland: A Romance of Many Dimensions*. London. Available at http://www.gutenberg.org/etext/201 or http://www.archive.org/details/flatlandromanceo00abbouoft.

Agarwal, S., et al. (2009). *Building Rome in a Day*. http://grail.cs.washington.edu/rome.

Amdahl, G. M., Blaauw, G. A., & Brooks, F. P. (1964). Architecture of the IBM System/360. *IBM Journal for Research and Development*, 8(2). (Reprinted in *IBM Journal for Research and Development*, 44(1/2), 2000).

Anderson, C. (2006). *The Long Tail: Why the Future of Business Is Selling Less of More*. Hyperion.

Annaud, J. (1986). *The Name of the Rose*.

Ariely, D. (2008). Keynote. *Authors@Google*, July 1. http://www.youtube.com/watch?v=VZv--sm9XXU.

Ariely, D. (2010). *The Upside of Irrationality: The Unexpected Benefits of Defying Logic at Work and at Home*. Harper.

Ashby, F. G., & Maddox, W. T. (2005). Human category learning. *Annual Review of Psychology*, 56, 149–178.

Ashby, F. G., & Spiering, B. J. (2004). The neurobiology of category learning. *Behavioral and Cognitive Neuroscience Reviews*, 3(2), 101–113.

Bachelard, G. (1969). *The Poetics of Space*. Beacon Press.

Balderston, D. (1986). *The Literary Universe of Jorge Luis Borges: An Index to References and Allusions to Persons, Titles. and Places in His Writings*. Greenwood Press.

Barabasi, A. (2003). *Linked*. Plume.

Baron, J. (2007). *Thinking and Deciding* (4th ed.). Cambridge University Press.

Barsalou, L. W. (1983). Ad hoc categories. *Memory and Cognition*, 11(3), 211–227.

Barsalou, L. W. (2003). Situated simulation in the human conceptual system. *Language and Cognitive Processes*, 18(5/6), 513–562.

Bates, M. (1989). The design of browsing and berrypicking techniques for the online search interface. *Online Review*, 13(October), 407–424. Available at http://gseis.ucla.edu/faculty/bates/berrypicking.html.

Bates, M. (2002). Toward an integrated model of information seeking and searching. *New Review of Information Behaviour Research*, 3, 1–15. Available at http://www.gseis.ucla.edu/faculty/bates/articles/info_SeekSearch-i-030329.html.

BBC. *Backstge: Open Data and Resources from the BBC*. http://backstage.bbc.co.uk.

Beesley, P., & Khan, O. (2009). *Responsive Architecture: Performing Instruments*. Situated Technologies Pamphlet 4. The Architectural League of New York. Available at http://www.situatedtechnologies.net/files/ST4-ResponsiveArchitecture.pdf.

Bellisario, G., & Ferrigno, C. (2010). User experience e televisione del futuro (User Experience and Future Television). In *Proceedings of the 4th Italian Information Architecture Summit*, Pisa, May 7-8. http://iasummit.architecta.org/2010/papers/ferrigno-bellisario.pdf.

Berlow, E. (2010). How complexity leads to simplicity. *TEDGlobal*. http://www.ted.com/talks/eric_berlow_how_complexity_leads_to_simplicity.html.

Bertolucci, K. (2003). Happiness is taxonomy: Four structures for Snoopy: Libraries' method of categorizing and classification. *Information Outlook*, March. http://findarticles.com/p/articles/mi_m0FWE/is_3_7/ai_99011617.

Bistagnino, L. (2009). *Design sistemico (System Design)*. Slow Food.

Blandford, A., & Attfield, S. (2010). *Interacting with Information*. Morgan and Claypool.

Boersma, P. (2004). *T-Model: Big IA is now UX. [BEEP]*. November 6. http://beep.peterboersma.com/2004/11/t-model-big-ia-is-now-ux.html.

Böhlen, M., & Frei, H. (2010). *MicroPublicPlaces*. Situated Technologies Pamphlet 6. The Architectural League of New York.

Bollnow, O. F. (1963). *Mensch und Raum (Man and Space)*. Kohlhammer.

Bolter, J. D., & Grusin, R. (2000). *Remediation: Understanding New Media*. MIT Press.

Bonsiepe, G. (1995). *Dall'oggetto all'interfaccia: Mutazioni del design*. Feltrinelli.

Borges, J. L. (1972). The analytical language of John Wilkins. In Borges, *Other Inquisitions: 1937-1952*. University of Texas Press. Available in a slightly different translation at http://www.crockford.com/wrrrld/wilkins.html.

Bosveld, J. (2009). The platypus genome is a mash-up of reptiles, birds, and mammals. *Discover Magazine*, (January). http://discovermagazine.com/2009/jan/090.

Bouchard, C. (2008). *The Hunt for Gollum*. Independent Online Cinema.

Bowker, G. C., & Star, S. L. (1999). *Sorting Things Out: Classification and Its Consequences*. The MIT Press.

Brabazon, T. (Ed.). (2008). *The Revolution Will Not Be Downloaded*. Chandon Publishing.

Brancheau, J. C., & Wetherbe, J. C. (1986). Information architectures: methods and practice. *Information Processing & Management*, 22(6), 453–463.

Brand, S. (1994). *How Buildings Learn*. Penguin Books.

Branit, B. (2009). *World Builder*. http://www.youtube.com/watch?v=VzFpg271sm8.

Briggs, A., & Burke, P. (2002). *A Social History of the Media*. Polity Press.

Bruno, G. (2007). *Public Intimacy: Architecture and the Visual Arts*. The MIT Press.

Bussolon, S. (2007). *La categorizzazione on line: aspetti teorici, metodologici ed applicativi*. (Online Categorization: Theoretical, Methodological and Practical Issues). PhD Dissertation. University of Trento. http://www.bussolon.it/dottorato/tesi.pdf.

Calori, C. (2007). *Signage and Wayfinding Design*. John Wiley & Sons.

Calvino, I. (1993). *Six Memos for the Next Millennium*. Vintage.

Cambiassi, R. (2006). BBC, prove tecniche di classificazione sociale (BBC A Testcase for Social Classification). *Apogeonline*, February 28. http://www.apogeonline.com/webzine/2006/02/28/01/200602280101.

Campbell, D. G., & Fast, K. (2006). From pace layering to resilience theory: The complex implications of tagging for information architecture. In *Proceedings of the 7th Information Architecture Summit*, Vancouver, March 23-27. ASIS&T. http://www.iasummit.org/2006/files/164_Presentation_Desc.pdf.

Carter, H. (1999). Information architecture. *Work Study*, 48(5), 182–185. http://dx.doi.org/10.1108/00438029910286026.

Case, D. O. (2005). Principle of least effort. In Fisher, et al. (2005, pp. 289–292).

Chazan, M. (2004). Locating Gesture: Leroi-Gourhan among the Cyborgs. http://www.semioticon .com/virtuals/Locating%20Gesture.pdf.

Child, T. (1887). A note on impressionist painting. *Harper's New Monthly Magazine*. http://harpers .org/archive/1887/01/0046953.

Ciolfi, L. (2004). Understanding spaces as places: extending interaction design paradigms. *Cognition, Technology & Work*, 6(1), 37–40.

Ciolfi, L., & Bannon, L. (2002). Designing interactive museum exhibits: Enhancing visitor curiosity through augmented artefacts. In Bagnara, et al. (Ed.), *Proceedings of ECCE11, European Conference on Cognitive Ergonomics*, Catania, September 2002. http://richie.idc.ul.ie/~luigina/ PapersPDFs/ECCE.pdf.

Cockburn, A., Gutwin, K., & Greenberg, S. (2007). A predictive model of menu performance. In *CHI '07: Proceedings of the SIGCHI Conference on Human Factors in Computing Systems*, San Jose, 28 April-3 May, 2007. ACM Press. http://portal.acm.org/citation.cfm?doid=1240624.1240723.

Cocuccioni, E. Il modello multimediale della mente (A Multimedia Model of the Mind). *La Critica*. *Available at* http://www.lacritica.net/selfspace/storyboarding/LQb.html.

Conan Doyle, A. (1995). *A Study in Scarlet*. http://www.gutenberg.org/ebooks/244.

Cortazar, J. (1987). *Hopscotch*. Pantheon.

Coward, L. A., & Salingaros, N. A. (2004). The Information architecture of cities. *Journal of Information Science*, 30(2), 107–118. Reprinted in Salingaros, N. A. (2005). *Principles of Urban Structure*. Techne Press. Available at http://zeta.math.utsa.edu/~yxk833/InfoCities. html.

Crockford, D. (2010). The Analytical Language of John Wilkins. *Douglas Crockford's Wrrrld Wide Web*. http://www.crockford.com/wrrrld/wilkins.html.

Crumlish, C. (2009). *Designing Social Interfaces: Principles, Patterns, and Practices for Improving the User Experience*. O'Reilly Media / Yahoo Press.

De Toni, A. F., & Comello, L. (2010). *Journey Into Complexity*. Available at http://www.lulu.com/ product/a-copertina-morbida/journey-into-complexity/11726755.

Denton, W. (2003). *How to Make a Faceted Classification and Put It On the Web*. Miskatonic University Press. November. http://www.miskatonic.org/library/facet-web-howto.html.

Denton, W. (2009). *Ranganathan's Prolegomena to Library Classification*. Miskatonic University Press. March 27. http://www.miskatonic.org/library/prolegomena.html.

Dictionary of Victorian London. http://www.victorianlondon.org.

Dillon, A. (2005). Opening keynote at EuroIA 2005. *1st European Information Architecture Summit*, Brussels, October 15-16. (See Gover 2006 for a synthesis).

Ding, W., & Lin, X. (2009). *Information Architecture: The Design and Integration of Information Spaces*. Morgan and Claypool.

Dini, A. (2006). I paradossi dell'imbarazzo della scelta (Paradoxes of Choice). *Nòva 24 - Il Sole 24 Ore*, September 7, 7.

Dourish, P. (1999). Following Where the Footprints Lead: Tracking Down New Roles for Social Navigation. In Munro et al. (Eds.). (1999, pp. 15–34).

Dourish, P., & Chalmers, M. (1994). *Running Out of Space: Models of Information Navigation*. Short paper presented at HCI '94, Glasgow. http://www.dcs.gla.ac.uk/~matthew/papers/ hci94.pdf.

Dourish, P., & Harrison, S. (1996). Re-Place-ing Space: The Roles of Place and Space in Collaborative Systems. In *CSCW '96: Proceedings of the 1996 ACM conference on Computer supported cooperative work* (pp. 67–76). ACM Press. http://doi.acm.org/10.1145/240080.240193.

Duffy, F. (1990). Measuring Building Performance. *Facilities*, 8(5), 17–20.

Dumézil, G. (1996). *Archaic Roman Religion*. The Johns Hopkins University Press.

East, R., Wright, M., & Vanhuele, M. (2008). *Consumer Behaviour: Applications in Marketing*. Sage.

Eco, U. (1997). *Kant and the Platypus: Essays on Language and Cognition*. Harcourt.

Eco, U. (2006). *The Name of the Rose*. Everyman's Library.

Eco, U. (2009). *Infinity of Lists*. Rizzoli International / Universal.

Egenter, N. (1992). Otto Friedrich Bollnow's anthropological concept of space. In *Proceedings of the 5th International Congress of the International Association for the Semiotics of Space*, Berlin, June 29-31. http://home.worldcom.ch/~negenter/012BollnowE1.html.

Evernden, R., & Evernden, E. (2003). *Information First: Integrating Knowledge and Information Architecture for Business Advantage*. Butterworth-Heinemann.

Falkheimer, J., & Jansson, A. (Eds.). (2006). *Geographies of Communication*. Nordicom.

Feldman, S. (2004). The high cost of not finding information. *KM World*, March 1. http://www.kmworld.com/Articles/ReadArticle.aspx?ArticleID=9534.

Fisher, K. E., Erdelez, S., & Mckechnie, L. (Eds.). (2005). *Theories of Information Behavior*. Information Today.

Foucault, M. (1994). *Order of Things: An Archeology of the Human Sciences*. Vintage.

Friedberg, A. (2006). *The Virtual Window: From Alberti to Microsoft*. The MIT Press.

Garrett, J. J. (2002). *The Elements of User Experience*. New Riders Publishing.

Garrett, J. J. (2009). *The Memphis Plenary*. http://www.jjg.net/ia/memphis/. (Transcript of the closing plenary address delivered March 22, 2009 at ASIS&T IA Summit 2009 in Memphis, Tenn).

Gnoli, C., Merli, G., Pavan, G., Bernuzzi, E., & Priano, M. (2008). Freely faceted classification for a Web-based bibliographic archive: The BioAcoustic Reference Database. In *Repositories of Knowledge in Digital Spaces: Accessibility, Sustainability, Semantic Interoperability*. 11th German ISKO conference, Konstanz, 20-22 February 2008. http://hdl.handle.net/10150/106114.

Gombrich, E. H. (2000). *Art and Illusion*. Bollingen.

Gombrich, E. H., Hochberg, J., & Black, M. (1973). *Art, Perception, and Reality*. The Johns Hopkins University Press.

Gotan Project. (2010). Rayuela (Hopscotch). In Gotan Project, *Tango 3.0* ¡Ya Basta! Records. http://www.youtube.com/watch?v=Zf6J-Hg_YWo.

Gould, S. J., & Vrba, E. S. (1982). Exaptation — a missing term in the science of form. *Paleobiology*, *8*(1), 4–15.

Gover, D. (2006). Euro IA Summit wrap-up. *Boxes and Arrows*, January 30. http://www.boxesandarrows.com/view/euro_ia_summit_wrap_up.

Greenaway, P. (1978). *A Walk Through H: The Reincarnation of an Ornithologist*. British Film Institute (BFI).

Greenfield, A. (2006). *Everyware: The Dawning Age of Ubiquitous Computing*. New Riders.

Greenfield, A. (2008). New day rising. *Adam Greenfield's Speedbird*, January 1. http://speedbird.wordpress.com/2008/01/01/new-day-rising/.

Greenfield, A., & Shepard, M. (2007). *Urban Computing and Its Discontents*. Situated Technologies Pamphlet 1. The Architectural League of New York. Also avaliable at http://www.situatedtechnologies.net/files/ST1-Urban_Computing.pdf.

Grossman, J. (2006). Designing for bridge experiences. *UX Matters*, June 30. http://www.uxmatters.com/mt/archives/2006/06/designing-for-bridge-experiences.php.

Haft, A. J. (1995). Maps, mazes, and monsters: The Iconography of the library in Umberto Eco's The Name of the Rose. *Studies in Iconography*, *14*, 9–50. Available in a slightly modified form at http://www.themodernword.com/eco/eco_papers_haft.html.

Hearst, M. A. (1996). Research in support of digital libraries at Xerox PARC: Part I: The changing social roles of documents. *D-Lib Magazine*, May. http://www.dlib.org/dlib/may96/05hearst.html.

Hearst, M. A. (2009). *Search User Interfaces*. Cambridge University Press. Available at http://searchuserinterfaces.com/book/.

Hempel, S. (2007). *The Strange Case of the Broad Street Pump: John Snow and the Mystery of Cholera*. University of California Press.

Hertzberger, H. (1971). Looking for the beach under the pavement. *Royal Institute of British Architects Journal*, *78*, 328–333.

Hill, S. (2000). An interview with Louis Rosenfeld and Peter Morville. *O'Reilly Media*, January 1. http://www.oreillynet.com/pub/a/oreilly/web/news/infoarch_0100.html.

Hinton, A. (2008). Linkosophy. In *Proceedings of the 9th Information Architecture Summit*, Miami, April 10-14. ASIS&T. http://www.slideshare.net/andrewhinton/linkosophy-355763.

Hinton, A. (2009). The machineries of context. *Journal of Information Architecture*, *1*(1). http://journalofia.org/volume1/issue1/04-hinton/jofia-0101-04-hinton.pdf.

Iacobelli, G. (Ed.). (2010). *Fashion Branding 3.0: La multicanalità come approccio strategico per il marketing della moda.* (Cross-channel as a Strategic Approach in Fashion Marketing). FrancoAngeli.

Inalhan, G., & Finch, E. (2004). Place attachment and sense of belonging. *Facilities*, *22*(5/6), 120–128.

Institute For The Future. (2000). The future of retail: Revitalizing bricks-and-mortar stores. *Institute for the Future Corporate Associates Program*, *11*(1).

Institute For The Future. (2004). Infrastructure for the New Geography. *Institute for the Future Technology Horizons Program, SR-869*, (August). http://www.iftf.org/system/files/deliverables/SR-869_Infrastructure_New_Geog.pdf.

Institute For The Future. (2006a). All the world's a game: The future of context-aware gaming. *Institute for the Future Technology Horizons Program, SR-997*, (May). http://www.iftf.org/system/files/deliverables/SR-997_Context_Aware_Gaming.pdf.

Institute For The Future. (2006b). The many faces of context awareness: A spectrum of technologies, applications, and impacts. *Institute for the Future Technology Horizons Program, SR-1014*, (September). http://www.iftf.org/system/files/deliverables/SR-1014_Many_Faces_Context_Awareness.pdf.

Institute For The Future. (2009). Blended reality: Superstructing reality, superstructing selves. *Institute for the Future Technology Horizons Program, SR-1221*, (February). http://www.iftf.org/system/files/deliverables/SR-122~2.pdf.

Iyengar, S. (2010). *The Art of Choosing*. Twelve.

Iyengar, S., & Lepper, M. R. (2000). When choice is demotivating: Can one desire too much of a good thing? *Journal of Personality and Social Psychology*, *79*, 995–1006. Available at http://www.columbia.edu/~ss957/articles/Choice:is_Demotivating.pdf.

Jainism Global Resource Center. *Elephant and the Blind Men*. http://www.jainworld.com/literature/story25.htm.

Jenkins, H. (2005). *Media Convergence*. http://web.mit.edu/cms/People/henry3/converge.html.

Jenkins, H. (2006). *Convergence Culture: Where Old and New Media Collide*. New York University Press.

Jenkins, H. (2009). The revenge of the origami unicorn: Seven principles of transmedia storytelling. *Confessions of an Aca/Fan*, December 12. http://henryjenkins.org/2009/12/the_revenge_of_the_origami_uni.html.

Jones, J. C. (1966). Designing methods reviewed. In Gregory (Ed.), *The Design Method*. Butterworths.

Jones, J. C. (1992). *Design Methods* (3rd ed.). John Wiley & Sons.

Kalbach, J. (2007). *Designing Web Navigation: Optimizing the User Experience*. O'Reilly Media.

Keane, M. T., & Eysenck, M. W. (2005). *Cognitive Psychology: A Student's Handbook*. Psychology Press.

Khan, O., Scholz, T., & Shepard, M. (2007-2010). *Situated Technologies Pamphlet Series*. Architectural League of New York. http://www.situatedtechnologies.net.

Klima, X. *Norsk Teknisk Museum*. http://www.tekniskmuseum.no/film-videoklipp/fra-utstillinger/klima-x.

Klyn, D. (2009). Conversation with Richard Saul Wurman. *Wildly Appropriate*. http://wildlyappropriate.com/?p=781.

Klyn, D. (2010). There is no such thing as Jesse James Garrett. *Wildly Appropriate*, March 19. http://wildlyappropriate.com/?p=570.

Knemeyer, D. (2004). Richard Saul Wurman: The InfoDesign interview. *InfoDesign*, (January). http://www.informationdesign.org/special/wurman_interview.htm.

Knights, A. J. (2007). The platypus. *All Empire*, July 24. http://www.allempires.com/article/index.php?q=The_Platypus.

Koch, T. (2005). *Cartographies of Disease: Maps, Mapping, and Medicine*. ESRI.

Kolson Hurley, A. (2010). I'm an Architect. *Architect Magazine*, July 8. http://www.architectmagazine.com/architects/im-an-architect.aspx.

Kopalle, P. K. (2010). Modeling retail phenomena. *Journal of Retailing, 86*(2), 117–124.

Kotler, N. G., Kotler, P., & Kotler, W. I. (2008). *Museum Marketing and Strategy: Designing Missions. Building Audiences, Generating Revenue and Resources*. Jossey-Bass.

Kourouthanassis, P. E., Giaglis, G. M., & Vrechopoulos, A. P. (2010). Enhancing user experience through pervasive information systems: The case of pervasive retailing. *International Journal of Information Management, 27*, 319–335.

Krug, S. (2005). *Don't Make Me Think: A Common Sense Approach to Web Usability*. New Riders Press.

Kruschke, J. K. (2005). Category learning. In Lamberts, & Goldstone (Eds.), *The Handbook of Cognition* (pp. 183–201). Sage Publications.

Kuhn, T. (1962). *The Structure of Scientific Revolutions*. University of Chicago Press.

Kuksov, D., & Villas-Boas, J. M. (2010). When more alternatives lead to less choice. *Marketing Science, 29*(3), 507–524.

Kuniawsky, M. (2010). *Smart Things: Ubiquitous Computing User Experience Design*. Morgan Kaufmann.

Kurowski, B. (2006). On the order of things. *Brandt.Kurowski.net*, October 29. http://web.archive.org/web/20071116055905/http://brandt.kurowski.net/2003/10/22/wilkins/.

Lakoff, G. (1987). *Women Fire and Dangerous Things: What Categories Reveal about the Mind*. University of Chicago Press.

Lakoff, G., & Johnson, M. (1980). *Metaphors We Live By*. University Of Chicago Press.

Lamantia, J. (2008). First fictions and the parable of the palace. *UX Matters*, November 3. http://www.uxmatters.com/mt/archives/2008/11/first-fictions-and-the-parable-of-the-palace.php.

Lamantia, J. (2009). Inside out: Interaction design for augmented reality. *UX Matters*, August 17. http://www.uxmatters.com/mt/archives/2009/08/inside-out-interaction-design-for-augmented-reality.php.

Lamantia, J. (2010). Playing well with others: Design principles for social augmented experiences. *UX Matters*, March 8. http://www.uxmatters.com/mt/archives/2010/03/playing-well-with-others-design-principles-for-social-augmented-experiences.php.

Lambe, P. (2007). *Organizing Knowledge: Taxonomies, Knowledge and Organization Effectiveness*. Chandos Publishing.

Landauer, T. K., & Nachbar, D. W. (1985). Selection from alphabetic and numeric menu tree using a touch screen: breadth, depth, and width. In *CHI '85 Proceedings of the SIGCHI Conference on Human Factors in Computing Systems* (pp. 73–78). ACM Press. http://dx.doi.org/10.1145/1165385.317470.

Langone, G. (2010). Redesign dell'architettura informativa degli Ospedali riuniti di Salerno (Information Architecture Redesign of the United Hospitals of Salerno). *Trovabile*, May 16. http://trovabile.org/articoli/architettura-informazione-ospedale.

Laurel, B. (1993). *Computers as Theatre*. Addison-Wesley.

Lawson, B. (2001). *Language of Space*. Architectural Press.

Lawson, B. (2005). *How Designers Think* (4th ed.). Architectural Press.

Layout. (2009–2010). *Mark-Up*. http://www.mark-up.it/articoli/0,1254, 41_ART_4506,00.html.

Lefebvre, M. (Ed.). (2006). *Landscape and Film*. Routledge.

Leganza, G. (2010). Topic overview: Information architecture. *Forrester Research*, January 21. http://www.forrester.com/rb/Research/topic_overview_information_architecture/q/id/55951/t/2.

Leroi-Ghouran, A. (1993). *Gesture and Speech*. MIT Press.

Lessig, L. (2008). *Remix: Making Art and Commerce Thrive in the Hybrid Economy*. Bloomsbury.

Levy, M., & Weitz, B. (2008). *Retailing Management*. McGraw-Hill/Irwin.

Lidwell, W., Holden, K., & Butler, J. (2003). *Universal Principles of Design*. Rockport Publishers.

Lipman, B. L. (2001). Why is language vague? Department of Economics, University of Wisconsin. http://citeseerx.ist.psu.edu/viewdoc/summary?doi=10.1.1.148.2778.

Lloyd, V. (2007). *Service Design*. TSO.

Loasby, K. (2006). Changing approaches to metadata at bbc.co.uk: From chaos to control and then letting go again. *Bulletin of the American Society for Information Science and Technology*, 33(1). October/November. http://www.asis.org/Bulletin/Oct-06/loasby.html.

Longhi, C. (2006). *Orlando furioso di Ariosto-Sanguineti per Luca Ronconi*. (The Frenzy of Orlando by Ariosto and Sanguineti for Luca Ronconi). Edizioni ETS.

Lynch, K. (1960). *The Image of the City*. The MIT Press.

Maeda, J. (2006). *The Laws of Simplicity*. The MIT Press.

Manguel, A., & Guadalupi, G. (1999). *The Dictionary of Imaginary Places*. Harcourt.

Manovich, L. (2002). *The Language of New Media*. MIT Press.

Marchionini, G. (2004). From information retrieval to information interaction. In McDonald, & Tait (Eds.), *Advances in Information Retrieval: 26th European Conference on IR Research* (pp. 1–11). Springer. ECIR 2004, Sunderland, UK, April 5-7, 2004, Proceedings. http://ils.unc.edu/~march/ECIR.pdf.

Marconi, I. (2010). Definisci la cosa. (Defining the Damn Thing). *Contest for The 4th Italian Information Architecture Summit*. Pisa, May 7-8. http://vimeo.com/11338696.

Maurer, D. See Spencer, D.

McCullough, M. (2004). *Digital Ground*. MIT Press.

McMullin, J., & Starmer, S. (2010). Leaving Flatland: Designing services and systems across channels. In *Proceedings of 11th Information Architecture Summit*, Phoenix, April 9-11. ASIS&T. http://2010.iasummit.org/talks/9702; http://www.slideshare.net/jessmcmullin/leaving-flatland-crosschannel-customer-experience-design.

Medin, D. L., & Rips, L. J. (2005). Concepts and categories: Memory, meaning, and metaphysics. In Holyoak (Ed.), *The Cambridge Handbook of Thinking and Reasoning* (pp. 37–72). Cambridge University Press.

Medin, D. L., & Schaffer, M. M. (1978). Context theory of classification learning. *Psychological Review, 85*(3), 207–238.

Merholtz, P. (2009). Desire lines. *Adaptive Path Blog*, October 27. http://www.adaptivepath.com/blog/2009/10/27/desire-lines-the-metaphor-that-keeps-on-giving.

Metro Group. *Future Store Initiative*. http://www.future-store.org; http://www.youtube.com/watch?v=j0k5_CQPx_U.

Microsoft Office Labs. *Retail Future Vision*. http://www.officelabs.com/projects/retailfuturevision/.

Minda, J. P., & Smith, J. D. (2001). Prototypes in category learning: The effects of category size, category structure, and stimulus complexity. *Journal of Experimental Psychology: Learning, Memory, and Cognition, 27*, 775–799.

Mistry, P. (2009a). *SixthSense: Integrating Information with the Real World*. http://www.pranavmistry.com/projects/sixthsense/.

Mistry, P. (2009b). *The Thrilling Potential of SixthSense Technology*. http://www.ted.com/talks/pranav_mistry_the_thrilling_potential_of_sixthsense_technology.html.

MIT. (2008). *Convergence Culture Consortium*. http://convergenceculture.org.

Mitchell, W. J. (1995). *City of Bits*. MIT Press.

Moggridge, B. (2007). *Designing Interactions*. The MIT Press.

Monaco, J. (1977). *How to Read a Film*. Oxford University Press.

Morin, E. (1990). *Introduction à la pensée complexe (Introduction to Complex Thought)*. ESF.

Morrogh, E. (2003). *Information Architecture, an Emerging 21st Century Profession*. Prentice Hall.

Morville, P. (2002a). Big architect, little architect. *Argus Center for Information Architecture*, July 27. http://argus-acia.com/strange_connections/strange004.html.

Morville, P. (2002b). Bottoms up: Designing complex, adaptive systems. *Dr Dobb's*. November 12, 2002. http://www.drdobbs.com/184411741.

Morville, P. (2004). A brief history of information architecture. In Gilchrist, & Mahon (Eds.), *Information Architecture: Designing Information Environments for Purpose* (p. xiii). Neal-Schuman Publishers.

Morville, P. (2005). *Ambient Findability*. O'Reilly Media.

Morville, P. (2010). Ubiquitous service design. *Semantic Studios*, April 19. http://semanticstudios.com/publications/semantics/000633.php.

Morville, P., & Callender, J. (2010). *Search Patterns: Design for Discovery*. O'Reilly Media.

Munro, A. J., Höök, K., & Benyon, D. (1999). *Social Navigation of Information Space*. Springer.

Norberg-Schulz, C. (1979). *Genius Loci: Towards a Phenomenology of Architecture*. Rizzoli.

Norman, D. (2009a). Systems thinking: A product is more than the product. *Interactions, 16*(5). September/October. http://interactions.acm.org/content/?p=1286.

Norman, D. (2009b). *The Design of Future Things*. Basic Books.

Norman, K. L. (1991). *The Psychology of Menu Selection: Designing Cognitive Control at the Human/Computer Interface*. Ablex Publishing. Available at http://www.lap.umd.edu/poms/.

Nosofsky, R. M. (1986). Attention, similarity, and the identification-categorization relationship. *Journal of Experimental Psychology: General, 115*(1), 39–57.

Nutt, D., King, A. L., Saulsbiry, W., & Blakemore, C. (2007). Development of a rational scale to assess the harm of drugs of potential misuse. *The Lancet, 369*. http://www.thelancet.com/journals/lancet/article/PIIS0140673607604644/abstract.

Olson, H. (2009). Social influences on classification. In Bates, & Maack (Eds.), *Encyclopedia of Library and Information Sciences* (3rd ed.). CRC Press.

O'Reilly, T., & Battelle, J. (2009). Web squared: Web 2.0 five years on. In *Proceedings of Web 2.0 Summit*, San Francisco. O'Reilly Media and TechWeb. October 20-22. http://www.web2summit .com/web2009/public/schedule/detail/10194.

Osservatorio Permanente Contenuti Digitali (2009). *Indagine 2009 (Survey 2009)*. http://www .osservatoriocontenutidigitali.it.

Pake, G. E. (1985). Research at Xerox PARC: a founder's assessment. *IEEE Spectrum, 22*(10), 54–61.

Pampaloni, L. (1971). Per un'analisi narrativa del "Furioso" (For a Narrative Analysis of Furioso). *Belfagor, 26,* 133–150.

Passini, R. (1984). *Wayfinding in Architecture*. Van Nostrand Reinhold Company.

Perdue, D. (2010). Dickens' London. *David Perdue's Charles Dickens Page*, http://charlesdickenspage .com/dickens_london.html.

Perscarmona, G. (2009). Il Flipper e la Nuvola: L'uso delle regole per passare dall'informazione alla conoscenza (The Pinball and the Cloud: Rules for Moving from Information to Knowledge). In *Proceedings of the 3rd Italian Information Architecture Summit*. Forlì, February 20-21. http:// www.viddler.com/explore/DElyMyth/videos/87/.

Persson, P. (2003). *Understanding Cinema*. Cambridge University Press.

Piccaluga, G. (1974). *Terminus: i segni di confine nella religione romana (Boundary Signs in Roman Religion)*. Edizioni dell'Ateneo.

Polanski, R. (2005). *Oliver Twist*. Sony Pictures.

Politano, A. (2006). *The Plan-less House*. Shinkenchiku Residential Design Competition.

Porter, E. S. (1903). *The Great Train Robbery*. American Memory from the Library of Congress. Available at http://memory.loc.gov/ammem/index.html.

Potente, D., & Salvini, E. (2009). Apple, IKEA and Their Integrated Architecture. *Bulletin of the American Society for Information Science and Technology, 35*(4), 32–42, April/May. http://www .asis.org/Bulletin/Apr-09/AprMay09_Potente-Salvini.pdf.

Quintarelli, E. (2005). Folksonomies: Power to the People. In *Proceedings of ISKO Italy Meeting*, Milan, June 24. http://www.iskoi.org/doc/folksonomies.htm.

Quintarelli, E., Resmini, A., & Rosati, L. (2008). The FaceTag engine: a semantic collaborative tagging tool. In Zambelli, et al. (Eds.), *Browsing Architecture: Metadata and Beyond* (pp. 204–217). Fraunhofer IRB.

Ranganathan, S. R. (1931). *The Five Laws of Library Science*. Madras Library Association. Available at http://hdl.handle.net/10150/105454.

Ranganathan, S. R. (1967). *Prolegomena to Library Classification*. Asia Publishing House. Available at http://hdl.handle.net/10150/106370.

Raskin, J. (2000). *The Humane Interface: New Directions for Designing Interactive Systems*. Addison-Wesley.

Raymond, E. S., & Landley, R. W. (2004). *The Art of Unix Usability*. Pearson Education. Available at http://www.catb.org/~esr/writings/taouu/html/index.html.

Reiss, E. (2010). In defense of "making it up as you go along." *Johnny Holland Magazine*, July 28. http://johnnyholland.org/2010/07/28/in-defense-of-making-it-up-as-you-go-along/.

Resmini, A. (2007). Del wayfinding, o del come se si entra finalmente nel labirinto, si hanno strane idee e, come accade nei labirinti, ci si perde (On Wayfinding). In Rosati (2007, pp. 115–136).

Resmini, A. (2010). Of patterns and structures. In *Andrea Resmini blog*. October 13. http:// andrearesmini.com/blog/of-patterns-and-structures.

Resmini, A., Byström, K., & Madsen, D. (2009). IA growing roots: Concerning the Journal of IA. *Bulletin of the American Society for Information Science and Technology*, *35*(3), 31–33. http://www .asis.org/Bulletin/Feb-09/FebMar09_Resmini_Bystrom_Madsen.pdf.

Resmini, A., & Rosati, L. (2007). From physical to digital environments (and back): Towards a cross-context information architecture. In *Proceedings of the 3rd European Information Architecture Summit*, Barcelona, September 21-22. http://www.euroia.org/2007/Programme .aspx.

Resmini, A., & Rosati, L. (2008). Semantic retail: Towards a cross-context information architecture. *Knowledge Organization*, *35*(1), 5–15.

Resmini, A., & Rosati, L. (2009). Information architecture for ubiquitous ecologies. In *MEDES '09 The International Conference on Management of Emergent Digital EcoSystems*, Lyon, October 27-30. http://doi.acm.org/10.1145/1643823.1643859. Available at http://andrearesmini.com/blog/ia-for-ubiquitous-ecologies.

Resmini, A., & Rosati, L. (2010). The semantic environment: Heuristics for a cross-context human-information interaction model. In Dubois, et al. (Eds.), *The Engineering of Mixed Reality Systems* (pp. 79–99). Springer.

Roam, D. (2009). American health care on (4) napkins. Now all together! *Digital Roam*, August 15. http://digitalroam.typepad.com/digital_roam/2009/08/american-health-care-on-4-napkins-now-all-together.html.

Robertson, R. (2009). *Mock-Epic Poetry from Pope to Heine*. Oxford University Press.

Ronda León, R. (2008). Arquitectura de Información: análisis histórico-conceptual (Information Architecture: A Historical-conceptual Analysis). *No Solo Usabilidad journal*, April 28. http://www.nosolousabilidad.com/articulos/historia_arquitectura_informacion.htm. Chronology available in English at http://www.rodrigoronda.com/rodriweb/ia_chronology.

Rosati, L. (2007). *Architettura dell'informazione: Trovabilità dagli oggetti quotidiani al web*. (Information Architecture: From Everyday Things to The Web) Apogeo.

Rosch, E. (1975). Cognitive reference points. *Cognitive Psychology*, *7*, 532–547.

Rosch, E., & Loyd, B. B. (1978). *Cognition and Categorization*. Erlbaum.

Rosen, R. (1999). *Essays on Life Itself*. Columbia University Press.

Rosenfeld, L., & Morville, P. (2006). *Information Architecture for the World Wide Web* (3rd ed.). O'Reilly Media.

Rubinelli L. (Ed.). (2009). *(Retail) Layout*. Mark-Up - Il Sole 24 Ore. http://www.mark-up.it/articoli/0,1254,41_ART_2241,00.html.

Samuelson, L., & Swinkels, J. (2006). Information, evolution and utility. *Theoretical Economics*, *1*, 119–142.

Saxe, J. G. (1873). The blind men and the elephant. In Saxe (Ed.), *The Poems of John Godfrey Saxe* (pp. 77–78). James R. Osgood and Company. Available at http://rack1.ul.cs.cmu.edu/is/saxe/.

Sbicca, C. (2009). *Musei, marketing e nuove tecnologie*. (Museums, Mareketing and New Technologies). Master Degree Dissertation. University for Foreigners of Perugia.

Schwartz, B. (2005). *The Paradox of Choice: Why More Is Less*. Harper Perennial.

Science Daily. (2005). UC Berkeley, French scientists find missing link between the whale and its closest relative, the hippo. *ScienceDaily*, February 7. http://www.sciencedaily.com/releases/2005/02/050205103109.htm.

SEMS (Search Engine Marketing Strategies). (2009). *Survey 2009*. http://www.sems.it/ricerche .htm.

Seow, S. C. (2005). Information theoretic models of HCI: A comparison of the Hick-Hyman Law and Fitts' Law. *Human-Computer Interaction*, *20*, 315–352. Available at http://citeseerx.ist.psu .edu/viewdoc/summary?doi=10.1.1.86.4509.

Shirky, C. (2005). Ontology is overrated. *Clay Shirky's Writings About the Internet*, http://www .shirky.com/writings/ontology_overrated.html.

Shirky, C. (2008). Gin, television, and social surplus. *Here Comes Everybody*, April 26. http://www .herecomeseverybody.org/2008/04/looking-for-the-mouse.html.

Simanek, D. E. (2006). *The Flat Earth*. http://www.lhup.edu/~dsimanek/flat/flateart.htm.

Smith, D. K., & Alexander, R. C. (1988). *Fumbling the Future: How Xerox Invented, Then Ignored, the First Personal Computer*. William Morrow & Co.

Smith, E. E., & Medin, D. L. (1981). *Categories and Concepts*. Harvard University Press.

Snavely, K. N. (2008). *Scene Reconstruction and Visualization from Internet Photo Collections*. PhD Dissertation. University of Washington. http://grail.cs.washington.edu/theses/SnavelyPhd .pdf.

Soddu, C., & Colabella, E. (1992). *Il progetto ambientale di morfogenesi*. (Environmental Morphogenetic Project). Leonardo.

Spence, R. (2001). *Information Visualization*. Pearsons.

Spencer, D. (2006a). Four modes of seeking information and how to design for them. *Boxes and Arrows*, March 14. http://www.boxesandarrows.com/view/four_modes_of_seeking_ information_and_how_to_design_for_them.

Spencer, D. (2006b). Lakoff's "Women, Fire and Dangerous Things": What every information architect should know. In *Proceedings of the 7th Information Architecture Summit*. Vancouver, March 23-27. ASIS&T. http://iasummit.org/2006/.

Stephenson, N. (1999). *Cryptonomicon*. Harper Perennial.

Stephenson, N. (2000). *Snow Crash*. Spectra.

Stephenson, N. (2003). *The Baroque Cycle*. William Morrow.

Sterling, B. (2005). *Shaping Things*. The MIT Press.

Streitz, N. A. (2005). From human-computer interaction to human-artifact interaction: Interaction design for smart environments. In Hemmje, et al. (Eds.), *From Integrated Publication and Information Systems to Information and Knowledge Environments: Essays Dedicated to Erich J. Neuhold*. Springer.

Summers, J. (1991). *Soho: A History of London's Most Colourful Neighborhood*. Bloomsbury.

Sunstein, C. R. (2008). *Infotopia*. Oxford University Press.

Tarantino, Q. (1994). *Pulp Fiction*. Miramax Films.

The John Snow Archive and Research Companion. http://johnsnow.matrix.msu.edu.

Tuan, Y. (1974). *Topophilia: A Study of Environmental Perception, Attitudes, and Value*. Prentice Hall.

Underhill, P. (2008). *Why We Buy: The Science of Shopping*. Simon & Schuster.

Van Dijck, P. (2003). *Information Architecture for Designers*. Rotovision.

Villella, A. F. (2000). Circular narratives: Highlights of popular cinema in the '90s. *Senses of Cinema*, 3. http://archive.sensesofcinema.com/contents/00/3/circular.html.

Vinten-Johansen, P., Brody, H., Paneth, N., Rachman, S., & Russell Rip, M. (2003). *Cholera, Chloroform and the Science of Medicine: A Life of John Snow*. Oxford University Press.

Walz, S. P. (2010). *Toward a Ludic Architecture: The Space of Play and Games*. ETC Press. Available at http://etc.cmu.edu/etcpress/content/toward-ludic-architecture-space-play-and-games.

Warne, K. (2007). Organization man. *Smithsonian Magazine*, (May). http://www.smithsonianmag. com/science-nature/tribute_linnaeus.html.

Weinberger, D. (2007). *Everything Is Miscellaneous: The Power of the New Digital Disorder*. Times Books.

Weiser, M. (1991). The computer for the 21st century. *Scientific American*, Special Issue on Communications, Computers, and Networks, September. http://www.ubiq.com/hypertext/ weiser/SciAmDraft3.html.

Weitzmann, L. M. (1995). *The Architecture of Information: Interpretation and Presentation of Information in Dynamic Environments*. PhD Dissertation. Massachusetts Institute of Technology. http://dspace.mit.edu/handle/1721.1/29085.

White, J. (2007). *London in the 19th Century*. Vintage Books.

Wikipedia (2010). Syntagmatic and paradigmatic relations. In Wikipedia (Ed.), *Course in General Linguistics*. http://en.wikipedia.org/wiki/Course_in_General_Linguistics.

Williamson, K. (2005). Ecological theory of human information behavior. In Fisher, et al. (2005 pp. 128–131).

Wodtke, C. (2002). *Information Architecture: Blueprints for the Web*. New Riders.

Wollen, P. (1997). *Signs and Meaning in the Cinema*. BFI Publishing.

Wurman, R. S. (1997). *Information Architects*. Graphis Inc.

Wurman, R. S. (2000). *Information Anxiety 2*. Que.

Wurman, R. S. (2010). Keynote at 11th Information Architecture Summit. In Parks, Richard Saul Wurman Keynote: A Podcast from the IA Summit 2010 in Phoenix, AZ. *Boxes and Arrows*, April 11. http://www.boxesandarrows.com/view/ia-summit-10-richard.

Xerox Parc Labs. (1995). *Ubiquitous Computing*. http://www.ubiq.com/hypertext/weiser/quicktime/UbiCompIntro.qt.

Index

Note: Page numbers followed by *f* indicates figure, *t* indicates table, *b* indicates box and *nb* indicates note.